GASES - CHARACTERISTICS, TYPES AND PROPERTIES

SILANE

CHEMISTRY, APPLICATIONS AND PERFORMANCE

GASES - CHARACTERISTICS, TYPES AND PROPERTIES

Additional books in this series can be found on Nova's website under the Series tab.

Additional E-books in this series can be found on Nova's website under the E-book tab.

GASES - CHARACTERISTICS, TYPES AND PROPERTIES

SILANE

CHEMISTRY, APPLICATIONS AND PERFORMANCE

KATSUO MORIGUCHI
AND
SUSUMU UTAGAWA
EDITORS

nova
publishers
New York

For permission to use material from this book please contact us:
Telephone 631-231-7269; Fax 631-231-8175
Web Site: http://www.novapublishers.com

Library of Congress Cataloging-in-Publication Data

Silane : chemistry, applications, and performance / [edited by] Katsuo Moriguchi and Susumu Utagawa.

pages cm

Includes bibliographical references and index.

ISBN 978-1-62257-432-2 (hardcover)

1. Silane. 2. Silane compounds. I. Moriguchi, Katsuo, editor of compilation. II. Utagawa, Susumu, editor of compilation.

QD181.S6S435 2012

546'.68322--dc23

2012024285

Published by Nova Science Publishers, Inc. † New York

CONTENTS

PREFACE

Silanes are synthetic compounds containing one or more Si-C bonds. They have wide application in industry, dentistry and medicine. They are used mainly as surface coating agents for surface modification and as coupling agents for adhesion promotion between two dissimilar materials. In this book, the authors present current research in the chemistry, applications and performance of silanes. Topics include the role silanes play in the processing of electronic chips and their thermo-chemical data; silane application in dental materials; thermal cationic polymerization of E-caprolactone using silanes; nanosilica surface modification by low-volatility silane coupling agents; and synthesis of a mesoporous silica hollow microsphere using methyltrimethoxy silane.

Chapter 1 - 2-Methacryloyloxyethyl phosphorylcholine (MPC) is a synthetic phosphorylcholine (PC) containing monomer which can polymerize or copolymerize with various vinyl monomers to synthesize MPC polymers having a wide variety of molecular architectures and functions. The MPC polymers have been widely used in surface modification of various substrate materials and devices to form cell-membrane-like surfaces, a kind of bioinspired biocompatible interfaces which have been shown to significantly resist nonspecific protein adsorption and cell adhesion. Among them, MPC copolymers incorporated with 3-methacryloxypropyl trimethoxysilane (MPTMSi) momoner unit, which is a typical methacrylate silane-coupling agent, have been specially designed for preparing biocompatible interfaces through a silanization process. In this chapter, we introduce cationic, neutral, and anionic MPC-MPTMSi copolymers and their applications for modification of glass-based substrates and devices, especially microfluidic devices which enable miniaturized and integrated chemical and biological applications.

Chapter 2 - In this chapter the status, trends, and challenges of working with silane and functional silanes are examined. The subject is far too vast to be covered in a single chapter, and much of what has been accomplished is proprietary and not available for review.

Thus the focus of this review is on standard practices and advancements published in the literature over the last twenty years; beginning with review of silane applications.Processes for formation and purification of silanes are reviewed with emphasis on the Direct, Siemens, Union Carbide/REC, Schmidt, and Ethyl/MEMC processes. The review continues with examination of the role silanes play in the processing of electronic chips, and analysis of hazards and public health issues one must be aware of when working with these compounds. The review concludes with examination of thermo-chemical data for silanes.

Chapter 3 - Silanes are synthetic compounds containing one or more Si-C bonds. They have wide application in industry, dentistry and medicine. They are used mainly as surface coating agents for surface modification and as coupling agents for adhesion promotion between two dissimilar materials. The role of silanes as either surface coating or coupling agents depends on their chemical structure. A wide variety of reactive organic functional groups can be bonded to the silicon atom such that silanes can react with the functional groups of many substrates. When silanes are used as a surface coating agent, the physical and chemical properties of the substrate surfaces are improved. As coupling agents, silanes enhance the chemical bonding between two dissimilar materials. Reaction mechanisms to illustrate these actions of silanes will be presented. Some techniques of surface treatment such as air-abrasion and chemical etching to enhance the adhesion between a silane and the substrate surfaces will also be covered. Findings from current studies of various silanes as surface-coating and coupling agents in different areas, with an emphasis on dentistry, will be discussed. The focus will be on the findings of performance of silanes as surface coating and coupling agents in different aspects of applications, including limitations in use. Finally, an outlook about on future developing trends of silanes in dental materials science will be provided. (219 words)

Chapter 4 - New initiating systems based on silanes and metal salts are proposed for an easy thermal cationic polymerization of ε-caprolactone at RT under air. They ensure good to excellent polymerizations. Moreover, a direct incorporation of the metal (Ag, Pd) through this one pot synthetic procedure is observed. The addition of small amounts of water or isopropanol allows a tuning of the polycaprolactone molecular weights. A coherent picture of the involved chemical mechanisms is presented.

Chapter 5 - Composites based on silicas and methyl metacrylate silane coupling agents are advanced materials for resin modification, thermoplastics, glass fiber reinforced epoxy resins, sensor technologies. The purpose of the present research is to combine silica-organic compounds with inorganic constituents to tailor the binding properties. It deals with a new method of covering silica nanoparticles by methyl methacrylate silanes in a pseudo-liquid state. The adsorption process and the chemical reactions of metacrylate-containing silanes and the surface of nanosized silica were examined by means of infrared spectroscopy, thermal analysis and transmission electron microscopy (TEM). The FTIR analysis confirmed the successful introduction of functional groups onto the nanosilica surface. The ratio of the parent free silanol and the grafted organosilyl groups was determined to be 1.4:1, i.e. 55% of the silane molecules react with the surface silanol groups in a monofunctional and 45% in a bifunctional way.

Chapter 6 - This chapter presents a new and facile method for the modification of alginate microcapsules using 3-aminopropyltriethoxysilane (APTES) in aqueous media. Spherical-shaped microcapsules (Alg-capsule) with carboxymethyl cellulose liquid core and calcium alginate shell were prepared by a conventional method, and then, Alg-capsules were modified to produce APTES/alginate hybrid microcapsule (AP-capsule) without using any organic solvents or precipitating agents. The optical observation showed that the texture of AP-capsules was more glossy and transparent than that of Alg-capsules. The surface morphology and elemental compositions of microcapsules were characterized by Scanning Electron Microscopy (SEM), Fourier Transform Infrared Spectroscopy (FT-IR) and X-ray Photoelectron Spectroscopy (XPS). The results suggest that APTES molecules were incorporated to the framework of the alginate shells via electrostatic interaction between amino groups of APTES and carboxyl groups of alginate, and the hybrid shells had a thin and homogeneous structure. The shrinking of the capsule shells and the accumulation of APTES in the capsule shells were observed during the reaction between Alg-capsules and APTES molecules. This observation suggests the successful introduction of APTES molecules to the capsule shells as well as the self-condensation of loaded APTES molecules. In addition, the shell thickness and the amount of APTES in the capsule shells were found to depend on not only the pH of reaction solution but also the concentrations of APTES. From the permeability tests of compounds with different molecular weight (methyl viologen, nicotinamide adenine dinucleotide, diaphorase, bovine serum albumin), it was observed that the protein with 24 kDa easily permeated through Alg-capsule shells but hardly through AP-capsule shells. Therefore, AP-capsules were believed to have a molecular weight cutoff point of less than 24 kDa.

To demonstrate a practical application of AP-capsules, formate dehydrogenase (FDH) was selected as a model enzyme, and then the enzyme was immobilized with the capsules. The encapsulation of FDH in AP-capsules exhibited higher recycling ability than that in Alg-capsules due to the suppression of enzyme leakage. These results indicate that the AP-capsule is effective as a support of enzyme immobilization.

Chapter 7 - A mesoporous silica hollow micro sphere was synthesized in presence of methyltrimethoxy silane for the first time. The synthesized material was characterized by various instrumental techniques such as small-angle X-ray scattering, scanning electron microscopy, transmission electron microscopy.

Thermogravimetry/Differential thermal analysis, Fourier transform Infrared spectroscopy and Nitrogen adsorption measurement. The analysis reveals the material is mesoporous and having hollow micosphere morphology.

In: Silane: Chemistry, Applications and Performance
Editors: Katsuo Moriguchi and Susumu Utagawa
ISBN: 978-1-62257-432-2

Chapter 1

MODIFICATION OF GLASS SUBSTRATES AND MICROFLUIDIC DEVICES BY SILANIZATION OF BIOINSPIRED 2-METHACRYLOYLOXYETHYL PHOSPHORYLCHOLINE (MPC) POLYMERS FOR PREPARING BIOCOMPATIBLE INTERFACES

Yan Xu[1,*], Madoka Takai[2] and Kazuhiko Ishihara[3]
[1]Nanoscience and Nanotechnology Research Center, Research Organization for the 21st Century, Osaka Prefecture University, Osaka, Japan
[2]Department of Bioengineering, School of Engineering, The University of Tokyo, Tokyo, Japan
[3]Department of Materials Engineering, School of Engineering, The University of Tokyo, Tokyo, Japan

ABSTRACT

2-Methacryloyloxyethyl phosphorylcholine (MPC) is a synthetic phosphorylcholine (PC) containing monomer which can polymerize or copolymerize with various vinyl monomers to synthesize MPC polymers having a wide variety of molecular architectures and functions. The MPC polymers have been widely used in surface modification of various substrate materials and devices to form cell-membrane-like surfaces, a kind of bioinspired biocompatible interfaces which have been shown to significantly resist nonspecific protein adsorption and cell adhesion. Among them, MPC copolymers incorporated with 3-methacryloxypropyl trimethoxysilane (MPTMSi) momoner unit, which is a typical methacrylate silane-coupling agent, have been specially designed for preparing biocompatible interfaces through a silanization process. In this chapter, we introduce cationic, neutral, and anionic MPC-MPTMSi copolymers and their applications for modification of glass-based substrates and devices, especially microfluidic devices which enable miniaturized and integrated chemical and biological applications.

* E-mail: y-xu@21c.osakafu-u.ac.jp.

1. SURFACE MODIFICATION OF SUBSTRATES OF MICROFLUIDIC DEVICES

1.1. Microfluidic Devices

Microfluidic devices, or the so-called lab-on-a-chip devices and micro-total analysis systems (μ-TAS), are systems that process or manipulate small amounts of fluids using channels with dimensions of typically sub-millimeter scale for miniaturized and integrated chemical and biological applications [1]. They offer the possibility for performing laboratory operations on a small (10^{-4} to 10^{-9} m) scale with considerably low (10^{-9} to 10^{-18} L) reagent consumption, fast response times, and better process control in comparison with their conventional laboratory counterparts [2-8]. They also offer the possibility of massive parallelization and integration of an entire laboratory on a single chip, thereby allowing samples to be analyzed at the point of need and diseases to be diagnosed at the point of care rather than at the conventional laboratories and hospitals [8-13]. Till now, microfluidic devices have become a distinct field with many appealing applications that could herald a new era of high integration and miniaturization in the fields of physics, chemistry, biology, medicine, pharmacy, and environmental science [1, 14-25].

1.2. Motivations of Surface Modification in Microfluidic Devices

The control of surface properties becomes an indispensable prerequisite for various applications of microfluidic devices. As the size decreases, the surface-area-to-volume ratio of a microfluidic device increases considerably. This certainly provides many unique merits in thermodynamics, reaction kinetics, and transport phenomena [1, 26, 27] but simultaneously magnifies the effect of nonspecific bindings, resulting in fluid anomalies and low performance [28-31]. This effect is especially unfavorable in microfluidic devices used in a broad range of biomedical applications ranging from immunoassays, DNA analysis, and cell diagnostics to highly integrated and miniaturized microfluidic systems, because nonspecific bioadsorption of proteins, DNAs, and cells degrades the assay sensitivity seriously.

Accordingly, the motivations for surface modification in microfluidic devices for successful biomedical applications can be classified into two groups: (i) reduce or control unwanted analyte-wall interactions, especially nonspecific protein adsorption and subsequent bioreactions; and (ii) construct versatile surfaces on a chip to realize particular functions, especially for example, control of electroosmotic flow (EOF) which is a state-of-the-art approach most widely applied in microfluidic devices for fluid delivering. To address these issues, different surface strategies using polymers [29, 32-42] have been adopted to modify the microchannels fabricated on substrate materials, including SiO_2-based materials [14, 15, 26, 43] such as glass and quartz, and synthetic polymers [44-48] such as poly(dimethylsiloxane) (PDMS), poly(methyl methacrylate) (PMMA), and poly(ethylene terephthalate) (PET).

1.3. Strategies and Materials for Surface Modification in Microfluidic Devices

In general, polymeric substrates are slightly hydrophobic, leading to difficulty in liquid introduction and notorious nonspecific analyte-wall adsorption. Therefore, surface modifications of polymer-based microfluidic devices are mainly driven by efforts to increase the hydrophilicity of the surface. Various covalent and non-covalent polymer modifications have been used for polymeric microfluidic devices through surface chemistry and physics [32, 35, 36, 38, 40-42]. Surface grafting of different hydrophilic monomers on PMMA substrates for microfluidic applications has been reported by Stachowiak *et al.* [49]. Results indicated that photo-grafted layers of PEG methacrylate (PEGMA) demonstrated the ability to substantially reduce nonspecific protein adsorption. Surface activation is commonly employed to oxidize and render hydrophilicity to PDMS surfaces. The reactive silanol functional groups generated by surface activation (for example, oxygen plasma treatments) can be used for subsequent surface chemistry such as silanization and various polymerizations. Hu *et al.* have grafted acrylic acid, acrylamide, 2-hydroxylethyl acrylate, dimediylactylamide, and PEG-monomethoxyl acrylate onto PDMS to yield hydrophilic surfaces [50]. The grafted surfaces exhibited greatly reduced adsorption of proteins and retained EOF mobility and stability upon exposure to air as compared to oxidized PDMS. Recently, Lin *et al.* have reported grafting of epoxy-modified hydrophilic polymers onto a PDMS microfluidic chip to resist nonspecific protein adsorption and suppress EOF [35].

For glass microfluidic devices, procedures for silanization of microchannels employing silane reagents in combination with different hydrophilic polymers have been commonly employed to suppress nonspecific protein adsorption and control EOF. Channel walls are covalently coated with hydrophilic polymers such as polyacrylamide, poly(ethylene glycol) (PEG), and poly(vinyl alcohol) (PVA) via a silane sublayer, which is first deposited on the wall by derivatization using silane reagents [29, 33, 34, 37, 39]. Surfaces with permanent, robust, and reproducible coverage have been achieved with these covalent polymer coatings, and simply-formed and regenerable modifications have been realized using polymers that get physically adsorbed to the microchannel wall. Physically adsorbed polymeric coatings can be prepared using neutral polymers such as PEG, PVA, and poly(N-hydroxyethylacrylamide) (PHEA) and their derivatives or using charged polymers such as polyethyleneimine (PEI) and chitosan. The use of charged polymeric coatings makes it possible to both control EOF and improve protein separation in chip-based electrophoresis.

Recently, in addition to the aforementioned conventional polymer surfaces, a polymer family containing 2-methacryloyloxyethyl phosphorylcholine (i.e., MPC polymers) have also attracted considerable attention for modifications of both glass and polymer-based microfluidic devices because of their high performance in the suppression of nonspecific adsorption of various proteins, platelets, and cells (Figure 1). MPC polymers have been used to form cell-membrane-like interfaces for chip applications *via* physical adsorption [51-58], chemical bonding [59-65], and photo-induced grafting [66, 67]. In this chapter, we focus on the development and applications of MPC polymer biocompatible interfaces for modification of glass substrates and microfluidic devices through silanization processes.

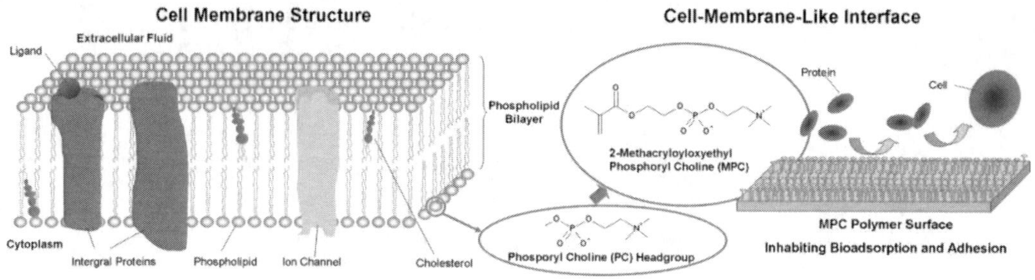

Figure 1. Schematics of cell membrane structure and cell-membrane-like surface with MPC polymer for inhibiting bioadsorption. (Image reproduced, with permission (Springer), from ref. 148).

2. Bioinspired 2-Methacryloyloxyethyl Phosphorylcholine (MPC) Polymers

The cell membrane is the biological and nanostructured membrane separating the intracellular components from the extracellular environment. It consists mainly of amphipathic phospholipids, spontaneously assembling as a continuous, spherical lipid bilayer structure where the hydrophilic head (polar) regions shield the hydrophobic tail regions from the surrounding polar fluid of the cytosolic and extracellular environments, and are consequently located at the outer faces of the resulting bilayer.[68] Since it serves diverse biological and physicochemical functions in cells, the cell membrane structure has inspired novel ideas in the fabrication of biomaterials and biointerfaces having cell-membrane-like structures (Figure 1). One of the major phospholipid polar groups on the cell membrane is phosphorylcholine (PC), which has a zwitterionic head group. The monomers comprising the PC group have been synthesized with attention to the chemical structure of the phospholipid molecules [69-71] and their polymers (referred to as phospholipid polymers) have been applied in the preparation of cell-membrane-like structures [72-78]. In 1990, Ishihara *et al.* reported the significant functions of phospholipid polymer materials in the biomedical field and successfully synthesized phospholipid monomers with high yield and excellent purity [79]. The representative monomer, MPC, is a methacrylate with a PC unit as a phospholipid polar group (Figure 1). Currently, the MPC monomer is industrially produced and distributed worldwide by a Japanese company. Many phospholipid polymers based on MPC polymer chemistry have been developed and studied for functionalized surface modification. The MPC monomer can copolymerize with various vinyl monomers to form phospholipid polymers having a wide variety of molecular architectures including random [53, 79, 80], block [81-83], graft [66, 67, 84], charged [59-61, 85], and end functional polymers [58, 62, 86-88]. They could form cell-membrane-like surfaces by coating the polymer [59-61, 89, 90], blending with the polymer [87, 88, 91, 92], and grafting to the polymer chains [90, 93, 94], and thereby provided biointerfaces capable of suppressing many biological responses such as nonspecific interactions with proteins, platelets, and cells (Figure 1). In addition, with incorporation of functional units enabling bioconjugation, these MPC polymers could also form PC-covered surfaces capable of selectively interacting with specific biomolecules [51, 52, 88, 95, 96]. Accordingly, the researches and applications cover biomembranes [97, 98], artificial organs [99, 100], vehicles for drug delivery systems (DDS) [88, 96, 101-104], scaffolds for tissue engineering [86, 88, 102, 105, 106], nanoparticles for immunoassays [95, 104], high sensitivity sensors [107, 108], and most recently microfluidic devices [51, 53, 54, 56, 59].

3. MPC Polymers Designed for Modification of Glass-Based Substrates and Microfluidic Devices through Silanization

3.1. Polymer Design

As a SiO_2-based material, glass is an initial and popular substrate to fabricate and develop microfluidic devices, especially for biological separation and analysis. This is not only because glass is chemically similar to fused silica and it is easy to transfer the classical separation behaviors from the conventional silica-based capillary electrophoresis (CE) to chip-based devices [109, 110], but also because their excellent properties such as physical rigidity, chemical inertness, optical transparency, thermal stability, and electrical insulation are very attractive and can satisfy some special chemical and analytical requisites, such as chemical syntheses, fluorescence detections, electrochemical analyses [43, 111]. As further downscaling of feature size to the nano scale, glass becomes an ideal substrate material to fabricate nanofluidic devices. This is because, in comparison with other popular substrate materials such as PDMS and other polymer-based materials, technologies for the fabrication of glass structures have already been well established in nanoscales with high resolution, reproducibility, and flexibility. Furthermore, the surface of glass is negatively charged and can generate a significant electroosmotic flow (EOF), which is favorable for some electrokinetic microfluidic devices [15, 112, 113].

The EOF is widely applied in miniaturized systems for fluid delivery [114, 115]. In some microfluidic systems with biological applications, proteins (or peptides) and cells are transported by the EOF. However, the irreversible, uncontrolled nonspecific protein adsorption and cell adhesion lead to low analytical efficiency and difficulty of EOF control, especially in the silica-based substrates, which make it particularly difficult to realize μ-TAS on glass for protein- and cell-related applications [33, 34, 37]. At the molecular level, rapid adsorption of proteins is generally thought as the initial stage of cell adhesion to a surface [77, 116, 117]. Therefore, prevention of protein adsorption is an indispensable prerequisite for successful biological applications in both protein- and cell-related microfluidic devices.

For surface modification of microchannels in microfluidic devices with polymers, several strategies such as filling coating [51, 53, 55, 59, 61], photo-induced polymerization [66, 67, 118], and electrospray deposition (ESD) [57, 58, 119, 120] have been developed. As an effective approach for suppression of protein adsorption, the strategy of surface coating has been widely applied. Till now, the coating materials for glass substrates and microchannels are mainly neutral, hydrophilic polymers, either covalently bound (such as polyamide and its derivatives), or physically adsorbed (such as dextran, poly (vinyl alcohol) and polyethylene oxide) [33, 34, 37]. They are effective for reduction of protein adsorption, but the EOF is usually dramatically reduced or totally eliminated simultaneously. For electrokinetically actuated microfluidic applications, it is unfavorable. In order to take advantage of EOF, charged polymer coatings are generally applied [121, 122]. However, the researches often showed that it is problematic, especially when charged proteins are handled on the surfaces with opposite charges, where the potential of adsorption may arise because of electrostatic attraction [123-125]. Apparently, it is quite difficult to construct a charged surface having ability of suppressing both anionic and cationic protein adsorptions.

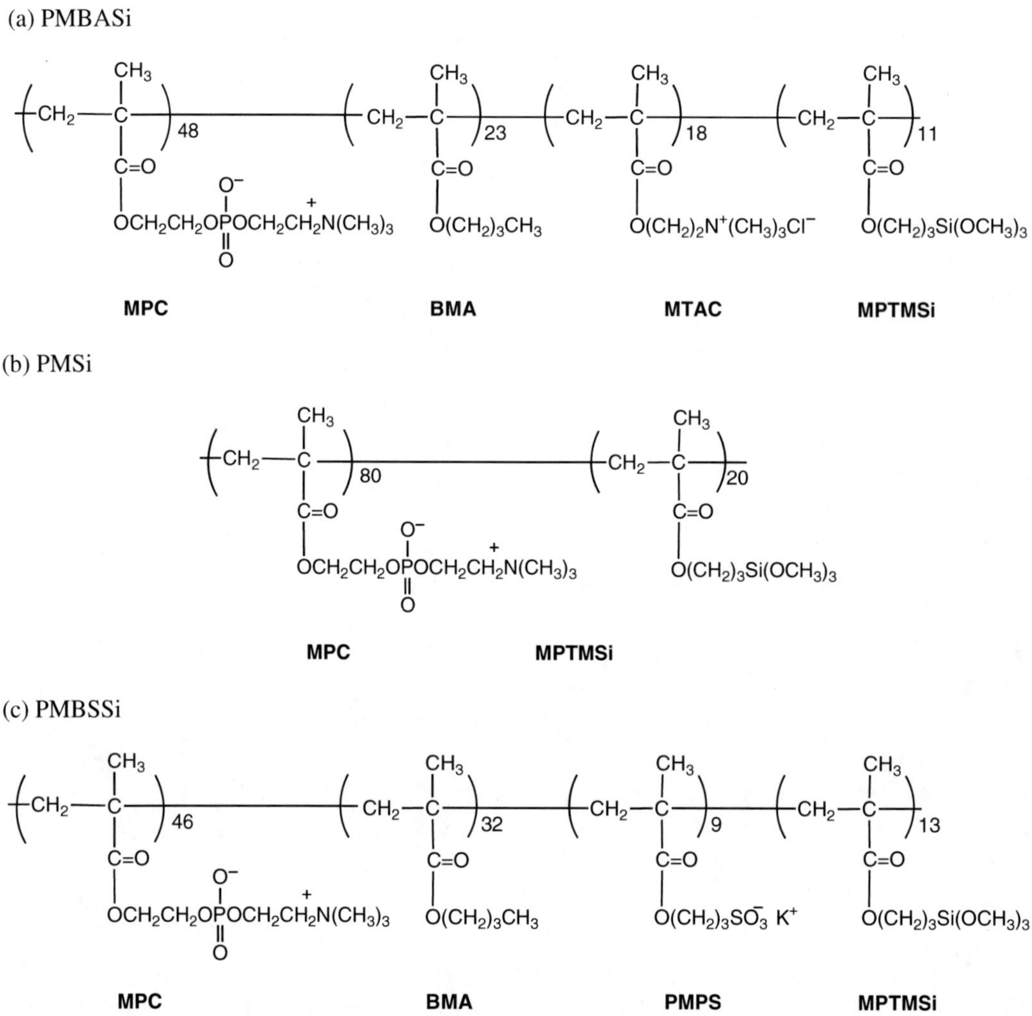

Figure 2. Chemical Structures of representative MPC polymers containing silane moieties and different charged moieties for glass microfluidic devices. (a) PMBASi (cationic), (b) PMSi (neutral) and (c) PMBSSi (anionic).

As introduced in the former section, the MPC polymer can be used to modify various surfaces, and their researches and applications cover wide areas from a bulk class to micro- and nano-scales accordingly. MPC polymers have been used in modification of substrates and microfluidic devices. For electrokinetic microfluidic devices on glass substrates, three representative MPC polymers containing silane moieties and different charged moieties were designed (Figure 2). They were poly(MPC-*co*-*n*-butyl methacrylate (BMA)-*co*-2-methacryloyloxyethyl trimethylammonium chloride (MTAC)-*co*-3-methacryloxypropyl trimethoxysilane (MPTMSi) (referred to as **PMBASi**, Figure 2a), poly(MPC-*co*-MPTMSi) (referred to as **PMSi**, Figure 2b), and poly(MPC-*co*-BMA-*co*-3-methacryloyloxypropyl sulfonate (PMPS)-*co*-MPTMSi) (referred to as **PMBSSi**, Figure 2c). They were designed to compose of different charged moieties (MTAC or PMPS) and a kind of silane coupling moiety MPTMSi, in consideration of suppressing protein adsorption, generating EOF, and stability of modification comprehensively. Polymer structures are schematically shown in Figure 2.

3.2. Polymer Synthesis and Bulk Properties

The three MPC copolymers—PMBASi, PMSi, and PMBSSi—can be synthesized through simple schemes by conventional radical polymerizations of their corresponding monomers. Briefly, the desired amounts of corresponding monomers, and AIBN (as initiator) were first dissolved in ethanol in a polymerization flask, and then the copolymerization was performed at 60 °C for 6 h in the flask after sealed. The polymerization product was collected after a reprecipitation from an ether/chloroform (7/3, v/v) mixture solvent. The remaining solvent was removed via a vacuum drying to get a white powder.

The structures and chemical compositions of three MPC copolymers were identified by ^{1}H NMR spectra (JEOL, 270 MHz, Tokyo, Japan):

Poly (MPC-*co*-BMA-*co*-MTAC-*co*-MPTMSi) (PMBASi, Figure 2a), (CD_3CD_2OD, δ): δ_H = 0.77–0.81 (–CH_2Si–), 0.90–1.09 (α-CH_3), 1.19–1.55 (–CH_3), 1.84–2.26 (–CH_2–), 3.00–3.17 (–N(CH_3)$_3$), 3.24–3.38 (–OCH_2CH_2OP), 3.69–4.44 (–OCH_2–).

Poly (MPC-*co*-MPTMSi) (PMSi, Figure 2b) , (D_2O, δ): δ_H = 0.60–0.65 (–CH_2Si–), 0.93–1.10 (α-CH_3), 1.14–1.46 (–CH_2–), 2.87–3.01 (–N(CH_3)$_3$), 3.54–3.70 (–OCH_2CH_2OP), 3.94–4.31 (–OCH_2–)).

Poly (MPC-*co*-BMA-*co*-PMPS-*co*-MPTMSi) (PMBSSi, Figure 2c) , (CD_3CD_2OD, δ): δ_H = 0.60–0.65 (–CH_2Si–), 0.90–1.04 (α-CH_3), 1.22–1.37 (–CH_3), 1.56–1.83 (–CH_2–), 3.15–3.24 (–N(CH_3)$_3$), 3.46–3.68 (–OCH_2CH_2OP), 3.97–4.24 (–OCH_2–).

Table 1. Molecular Properties of PMBASi, PMSi, and PMBSSi. (Table reproduced, with permission (Elsevier), from ref. 61.)

Abb.	Composition (mole fraction)[a] [MPC/BMA/X[b]/MPTMSi]	Molecular weight (*Mw*)[c]	Polydispersity ratio(*Mw/Mn*)[d]	Solubility[e]	
				H_2O	EtOH
PMBASi	48/23/18/11	9.9 × 10^3	1.1	+	+
PMSi	80/0/0/20	9.7 × 10^3	1.1	+	+
PMBSSi	46/32/ 9/13	11.8 × 10^3	1.2	+	+

[a] Determined by ^{1}H NMR.
[b] X is MTAC in PMBASi or PMPS in PMBSSi.
[c] Weight average molecular weight(*Mw*);determined by GPC, PEO standard.
[d] *Mn*: number average molecular weight.
[e] Evaluated by dissolving 0.1 g polymer powder in a 10 mL solvent; +: soluble.

Table 2. Atomic concentrations of different sample surfaces determined by XPS measurements. (Table reproduced, with permission (Elsevier), from ref. 61.)

Sample surface	Atomic concentration (%)					
	C_{1s}	O_{1s}	N_{1s}	P_{2p}	S_{2p}	Si_{2p}
Glass	8.9	60.7	n.d.	n.d.	n.d.	30.4
PMBASi	62.8	28.1	4.5	3.2	n.d.	1.4
PMSi	60.1	28.7	4.8	4.8	n.d.	1.6
PMBSSi	61.9	28.9	3.7	3.8	0.5	1.2

n.d.: not detected. The concentrations of polymer solutions for coatings are 3.00 mg mL^{-1} (*high-density coatings*).

The molecular properties of three MPC polymers are listed in Table 1. Besides MPC, all three copolymers contain MPTMSi which is a typical methacrylate silane-coupling agent. Silane-coupling moieties such as the MPTMSi moieties in a copolymer are very useful in many applications because they can react with various inorganic materials such as glass, silica, silicone, and metals to form a chemical bond with surface or can induce a cross-linkable coating on various organic materials such as poly(ethylene terephthalate) (PET), poly(vinyl chloride) (PVC), etc. [126]. The compositions of MPTMSi in all three copolymers are higher than 10%, a composition which has been proved sufficient for a MPC copolymer to form a stable covalent coating on a glass substrate [60]. The MTAC moieties of PMBASi and the PMPS moieties of PMBSSi are two types of methacrylate units respectively having oppositely charged groups. Furthermore, PMBASi and PMBSSi share another common feature as they contain almost same amounts of the MPC fraction. As shown in Table 1, all these three MPC polymers have similar molecular weights, possess narrow polydispersities, and are soluble in both water and ethanol. The solubility of three polymers was evaluated using several common solvents. The result reveals that three polymers have similar solubility in these solvents. They all dissolve in water, methanol and ethanol, but do not easily dissolve in the other solvents such as acetone, ether and IPA (Table 1).

4. Interfacial Properties of Glass Substrates Silanized with MPC Polymers

4.1. Surface Modification Process

As discussed in *Section 3*, the MPC polymers (PMBASi, PMSi and PMBSSi) are dissolved in ethanol, which is favorable in the modification of microchannel. Thus, the polymer alcoholic solutions were applied to modify substrates and no other additives were added, which provides good properties of easy-handling, quick-treatment and nearly no solvent remnant after treatment. The modification process for plate substrates is very simple, only one-step dip coating. Theoretically, as long as the glass surface is adequately in contact with the polymer solution, effective modification can be accomplished owing to the MPTMSi units in the polymers, which are silane coupling agents and can be chemically bonded with glass substrates. To characterize the interfacial properties of these copolymers, glass substrates respectively with three copolymer covalent coatings were prepared according to the protocol. Including uncoated, PMBASi-coated, PMSi-coated, and PMBSSi-coated glass substrates, four sample surfaces were used in the subsequent investigations. Figure 3 illustrates the differences between the four sample surfaces with highlighting the variation in the surface charge.

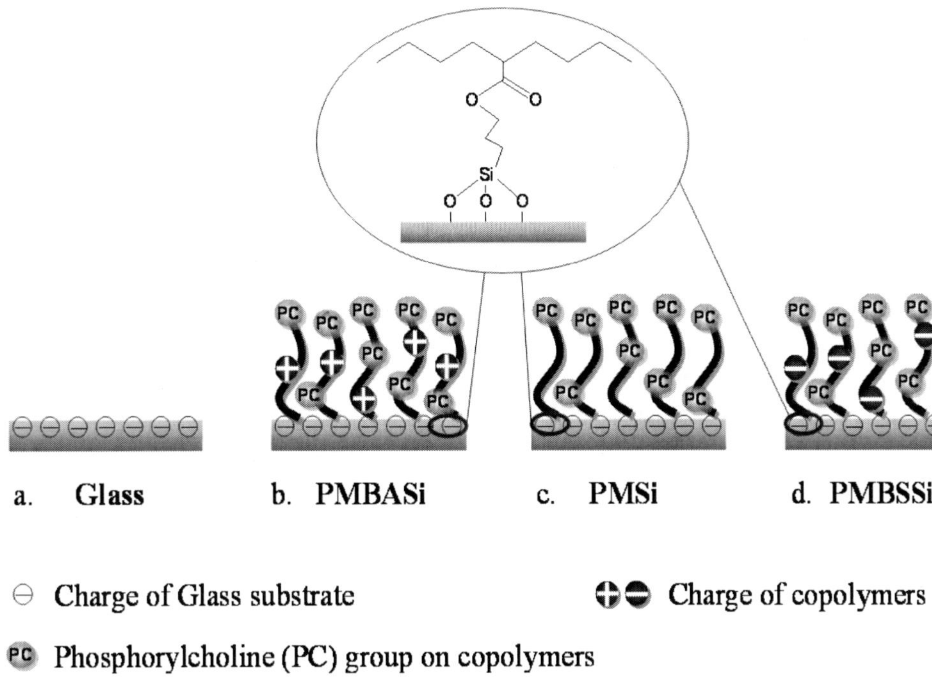

Figure 3. Glass substrates that are modified using different MPC copolymers and exhibit different surface charges. (a) Uncoated glass; (b) PMBASi-coated glass substrate; (c) PMSi-coated glass substrate; (d) PMBSSi-coated glass substrate. The inset demonstrates the chemical bond between the glass substrate and the MPC polymers. (Image reproduced, with permission (Elsevier), from ref. 61).

4.2. Interfacial Properties

4.2.1. Surface Elemental Composition

The surface elemental composition of the sample was determined by X-ray photoelectron spectroscopy (XPS; Axis-His, Shimadzu/KRATOS, Kyoto, Japan). Their spectra of nitrogen (N_{1s}), phosphorus (P_{2p}), and silicon (Si_{2p}) are shown in Figure 4. The XPS data of atomic concentrations of different sample surfaces are listed in Table 2. Those three MPC copolymers coated substrates were compared with the uncoated substrates. XPS signals at 133 eV attributed to the phosphorus 2p peak and at 402 eV attributed to the nitrogen 1s peak were detected on the PMBASi-, PMSi-, and PMBSSi-coated substrates but not on the uncoated substrates (Figure 4), revealing the existence of MPC moieties on these coated substrates. On both the PMBSSi- and the PMSi-coated surfaces, the atomic concentrations of nitrogen and phosphorus were almost same, while on the PMBASi-coated surface, the nitrogen concentration was quite higher than the phosphorus concentration due to the trimethylammonium groups of the MTAC moieties on PMBASi (Table 2). The sulfur 2p peak

(BE ~ 168 eV) was only detected on the PMBSSi-coated surface, owing to the sulfonate groups of PMPS moieties on PMBSSi. Furthermore, signals at 102 eV attributed to the silicon 2p peak were detected on all samples regardless of whether or not polymer coatings were applied (Figure 4). However, the atomic concentrations of silicon on the MPC copolymers coated substrates were greatly lower than those on the uncoated substrate (Table 2), suggesting that the substrates were effectively covered by the coating. These results indicate that the substrates were successfully modified by these MPC copolymers.

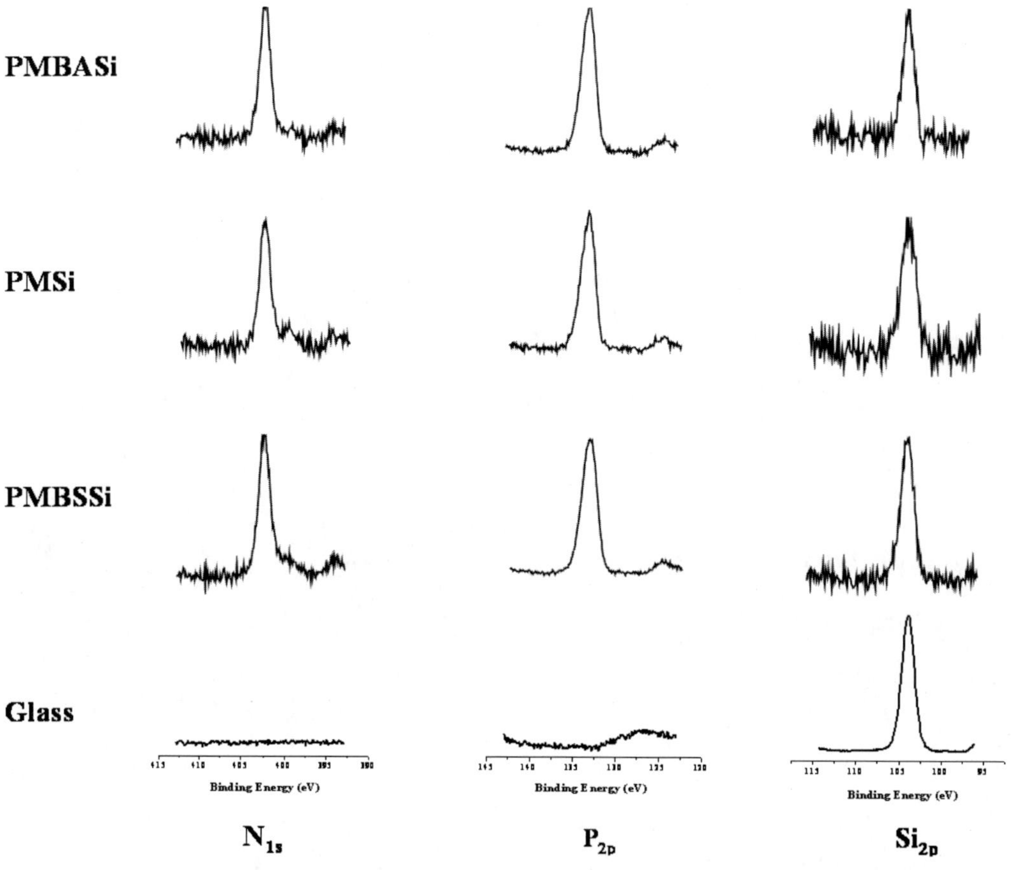

Figure 4. XPS spectra of nitrogen (N_{1s}), phosphorus (P_{2p}), and silicon (Si_{2p}) on glass substrates coated with PMBASi, PMSi, and PMBSSi in comparison with those on uncoated glass. (Image reproduced, with permission (Elsevier), from ref. 61).

4.2.2. Surface Density

The density and thickness of an MPC polymer interface can be controlled by adjusting parameters such as the concentration of the modification solution. Figure 5 shows the XPS spectra of nitrogen (N_{1s}) and phosphorous (P_{2p}) on the high-concentration (3.00 mg mL^{-1}) and low-concentration (0.30 mg mL^{-1}) PMBSSi coated glass substrate surfaces. The relative peak intensities of both elements were decreased with increasing the concentration of PMBSSi, implying that comparing with the high-concentration coating, the low-concentration coating might make more amount of PMBSSi reacted on the substrate. Figure 6 demonstrates the relationship between the PMBSSi coating concentration and coating density. The coating density was determined with a quartz crystal microbalance (QCM; Q-Sense D300 system, Q-

Sense AB, Goteborg, Sweden). In the figure, *Non-coating* (density = 0.00 μg cm^{-2}), *Low-density coating* (density = 1.19 μg cm^{-2}) and *High-density coating* (density = 11.14 μg cm^{-2}) represent the uncoated, low-concentration (0.30 mg mL^{-1}) PMBSSi coated, and high-concentration (3.00 mg mL^{-1}) PMBSSi coated glass surfaces, respectively. A linear fit is applicable to describe the relationship between the coating concentration and the coating density, which reveals that the coating density can be adjusted by changing the coating concentration of the polymer solution [60].

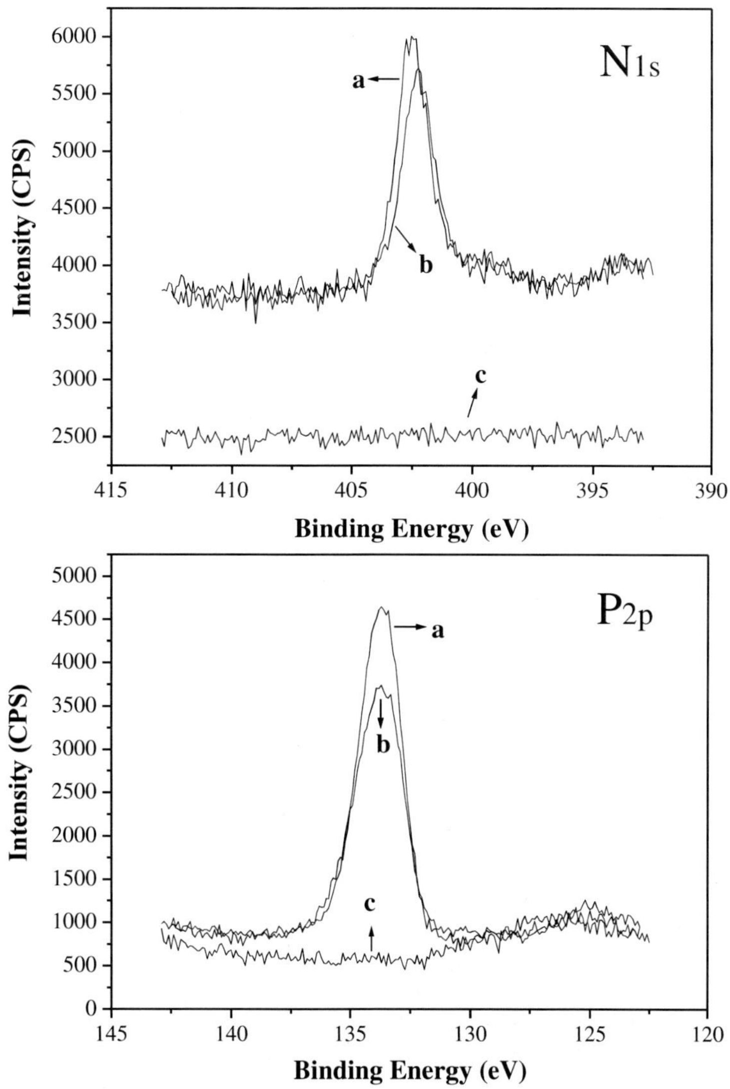

Figure 5. XPS spectra of nitrogen (N_{1s}) and phosphorus (P_{2p}) on different sample surfaces: (a) glass substrate surface with a high-concentration (3.00 mg mL^{-1}) PMBSSi coating (i.e., *high-density coating* sample); (b) glass substrate surface with a low-concentration (0.30 mg mL^{-1}) PMBSSi coating (i.e., *low-density coating* sample); (c) uncoated glass substrate surface (i.e., non-coating sample). The spectra were recorded using a take-off angle of 90° with respect to the sample surface. (Image reproduced, with permission (ACS), from ref. 60).

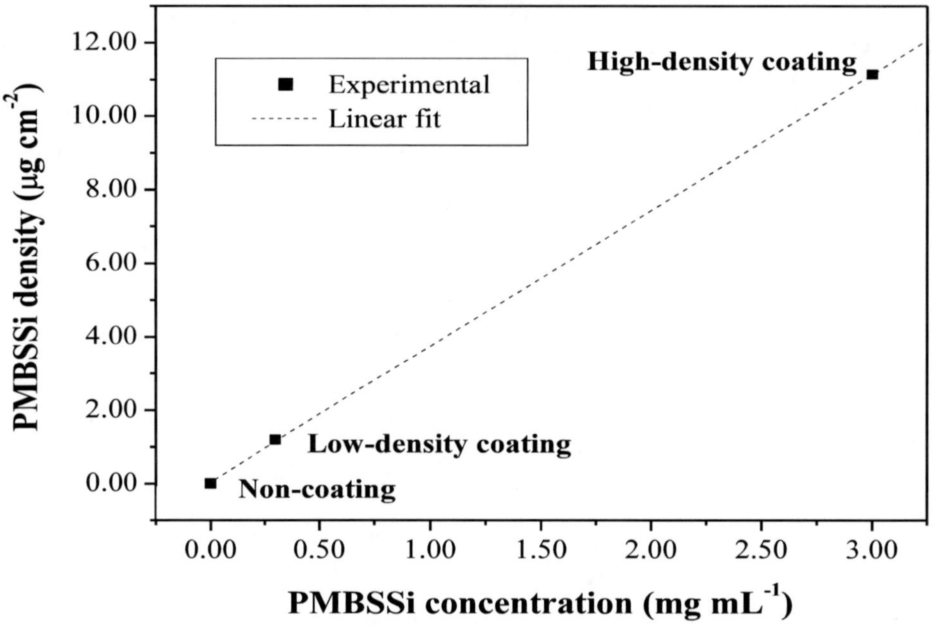

Figure 6. PMBSSi densities on sample surfaces *versus* PMBSSi concentrations for different coating densities. The linear fit ($y = 0.04 + 3.70x$, $R^2 = 1.00$) is used to describe the relationship between the PMBSSi coating density and the PMBSSi concentration for different coating densities. (Image reproduced, with permission (ACS), from ref. 60).

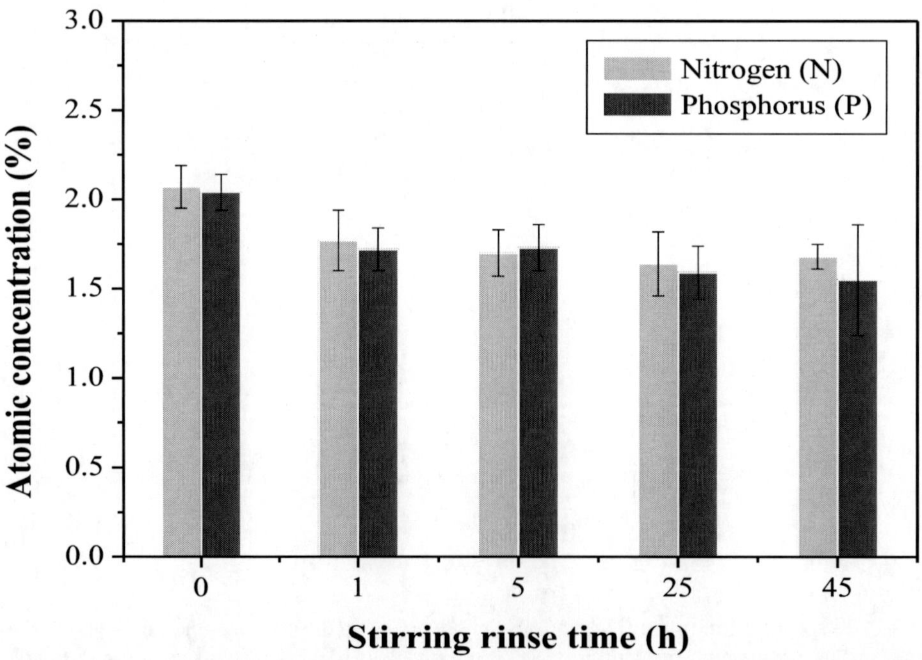

Figure 7. Changes in atomic concentration of nitrogen and phosphorus (determined by XPS) on glass substrate surfaces with PMBSSi coatings during rinsing by stirring at 300 rpm in distilled water for 45 h. Data are mean ± SD, n = 3. (Image reproduced, with permission (ACS), from ref. 60) .

4.2.3. Surface Stability

The coating stability was examined in distilled water undergoing a relatively high speed of magnetic stirring (300 rpm). The surface atomic concentrations of nitrogen (N) and phosphorous (P) on the glass surfaces with PMBSSi coatings, which were determined by XPS, remained almost constant during the entire 45-h rinse process (Figure 7). Thus, the PMBSSi coating layer could not be detached by fluidic rinsing. This reveals that the coating can effectively function for a long time even under long-term fluidic working conditions owing to the chemical bonding between the substrate and the polymer layer via a silanization process.

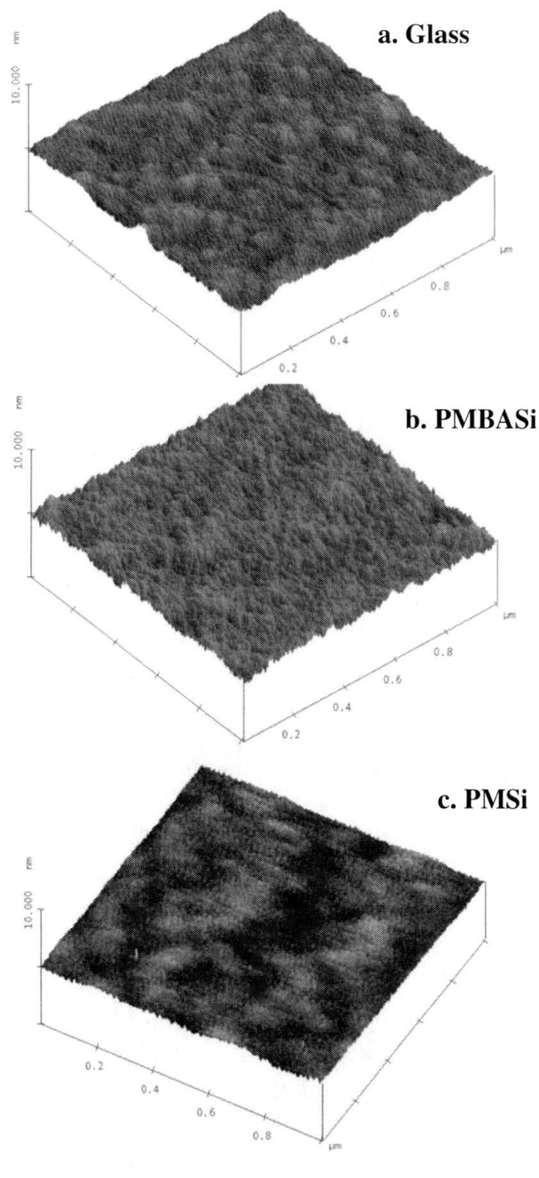

Figure 8. (Continued).

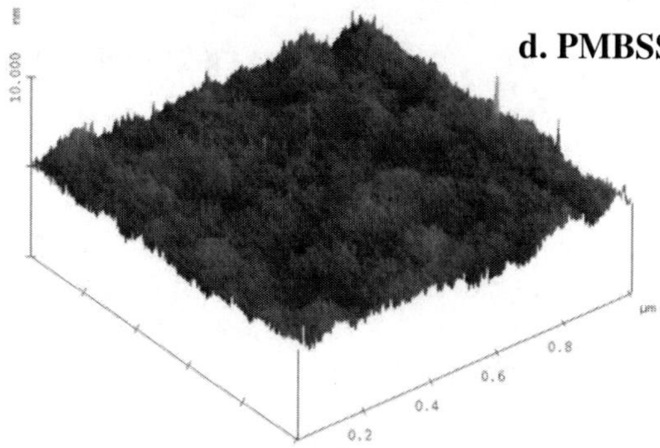

Figure 8. AFM images (1.0 µm × 1.0 µm areas) of different sample surfaces. (a) Uncoated glass substrate; (b) PMBASi-coated glass substrate; (c) PMSi-coated glass substrate; (d) PMBSSi-coated glass substrate. Polymer concentration was 3.00 mg mL^{-1}, i.e., *high-density coatings*. (Image reproduced, with permission (Elsevier), from ref. 61).

4.2.4. Surface Morphology

The surface topographies before and after coating were characterized by atomic force microscopy (AFM; Bioscope, Nanoscope IIIa, Veeco, Santa Barbara, CA). The topography images (Figure 8) demonstrate that all three coated surfaces exhibited as smooth as the uncoated surfaces. The smoothness of sample surfaces was quantitatively verified by analyzing their root-mean-square (rms) roughness. The uncoated substrate had a very low rms value of 0.43 ± 0.27 nm. The PMBASi-, PMSi-, and PMBSSi-coated surface exhibited rms values of 0.36 ± 0.19 nm, 0.15 ± 0.03 nm, and 0.37 ± 0.14 nm, respectively. The rms roughness values of the three coated surfaces were approximate to or lower than that of the uncoated surface, indicating that these coatings did not induce the increase in the surface roughness. Especially, the cationic PMBASi and the anionic PMBSSi-coated surfaces had almost equal rms roughness values, suggesting the type of the charged moiety may not affect the surface roughness.

4.2.5. Surface Wettability

To investigate the surface wettability, both the static contact angle (SCA) analysis and the dynamic contact angle (DCA) analysis were performed. The uncoated glass substrate surface was hydrophilic as it had a 31.2° static contact angle. When a PMSi coating was applied, the average contact angle decreased to 10.5°, indicating a highly hydrophilic surface. Both static contact angles of the PMBASi- and the PMBSSi-coated surfaces were lower than that of the uncoated glass surface but higher than that of the PMSi-coated surface. This reveals that, on the one hand both the PMBSASi and the PMBSSi coatings were quite hydrophilic in general, and on the other hand the hydrophobic moieties (i.e., BMA) on both coatings were likely to affect the surface wettability. The role of BMA moieties in the surface wettability was further elucidated by the DCA analysis. The DCA measurement is an easy method to determine the advancing and receding contact angles which are parameters of the surface wettability. The advancing contact angle, θ_A, is sensitive to the hydrophobic surface component, and the receding contact angle, θ_R, is sensitive to the hydrophilic surface component. The DCA profile provides dynamic information on the process from a dry state to a wet state of a surface when in contact with an aqueous medium. Figure 9 shows the DCA profiles of different sample surfaces. Table 3 summarizes the values of θ_A and θ_R obtained from Figure 9. As shown in Table 3, in contrast to the PMSi-coated surface, which displayed a low θ_A and a supper low θ_R corresponding to a small hysteresis (Figure 9), both the PMBASi- and the PMBSSi-coated surfaces had quite high θ_A and low θ_R, resulting in large hystereses (Figure 9). Under dry conditions the phosphorylcholine groups of PMBASi and PMBSSi were covered with the hydrophobic moieties of polymer chains to minimize the surface free energy, while the hydrophilic moieties (especially phosphorylcholine groups) became to expose to the aqueous environment to reduce the interfacial energy when contact with water. Hence, the observed large hystereses of PMBASi and PMBSSi may be related to the reorientation of the phosphorylcholine groups when the surface changed from a dry state to a wet state. Through a surface reorientation process, the phosphorylcholine groups would become dominate the outmost surface and thereby the hydrophobic moieties would be buried. This is believed very important for a MPC copolymer to play their functions in aqueous mediums [127, 128]. According to the value of mobility factor (MF), both the PMBASi- and the PMBSSi-coated surfaces had high mobilities of the polymer chains, implying that when in contact with an aqueous environment, surface reorientation may easily occur in both PMBASi- and PMBSSi-coated surfaces. In addition, a very small hysteresis was observed in the DCA profile of the PMSi-coated surface, which is possibly because phosphorylcholine groups enriched the polymer network and abundantly existed at the surface even under dry conditions [129]. Therefore, the nature of the observed difference in the dynamic wetting process between the PMSi surface and the PMBASi or PMBSSi surface should be ascribed to the introduction of the BMA moieties, which contributed the hydrophobic surface component and increased the surface hydrophobic/hydrohphlic heterogeneity in the coatings.

Table 3. Advancing and receding contact angles of different sample surfaces. (Table reproduced, with permission (Elsevier), from ref. 61.)

Sample surface	Contact angle (degree)			
	Advancing	Receding	Hysteresis	Mobility Factor
Glass	37.1	19.3	17.8	0.48
PMBASi	72.0	11.3	60.7	0.84
PMSi	22.3	0.0	22.3	1.00
PMBSSi	84.4	15.9	68.5	0.81

The concentrations of polymer solutions for coatings are 3.00 mg mL^{-1} (*high-density coatings*).

Table 4. Surface Properties of glass substrates with different PMBSSi coating densities. (Table reproduced, with permission (ACS), from ref. 60.)

Sample	Density[a] (μg cm^{-2})	Contact Angle[b] (°)	rms Roughness (nm)		ζ-potential[c] (mV)
			Dry	Wet	
Non-coating	0.00	31.2 ± 1.2	0.51	0.60	−47.4 ± 0.9
Low-density coating	1.19 ± 0.01	23.8 ± 2.8	0.30	0.79	−14.3 ± 1.9
High-density coating	11.14 ± 0.05	13.4 ± 1.3	0.32	0.69	−24.2 ± 2.5

[a] Data are mean ± SD, n =3; [b] Data are mean ± SD, n ≥ 6; [c] Data are mean ± SD, n = 6.

4.2.6. Surface ζ-Potentials

The coatings not only changed the surface elemental composition but also altered the surface charge of the original substrate. After coating, the charged moieties of the copolymer played their functions in the form of the interface. The surface ζ-potentials of sample surfaces were measured to characterize their surface charge in the aqueous environment. As shown in Figure 10, after coated with the neutral MPC copolymer PMSi, the ζ-potential of the substrate changed from an original value of −47.4 ± 0.9 mV to a value of near 0 mV; the cationic PMBASi-coated substrates and the anionic PMBSSi-coated substrates exhibited ζ-potentials of 26.1 ± 0.6 mV and −24.2 ± 2.5 mV, respectively; the signs of the ζ-potential values were finely in agreement with the variation in the charge properties of the corresponding copolymers; the absolute values of two ζ-potentials of cationic PMBASi and anionic PMBSSi coatings were approximately equal, making them quite comparable. These features indicate that after coating, the charge of the original substrate was shielded, and as a result the polymer layer instead of the surface of the original substrate became dominate the surface charge condition.

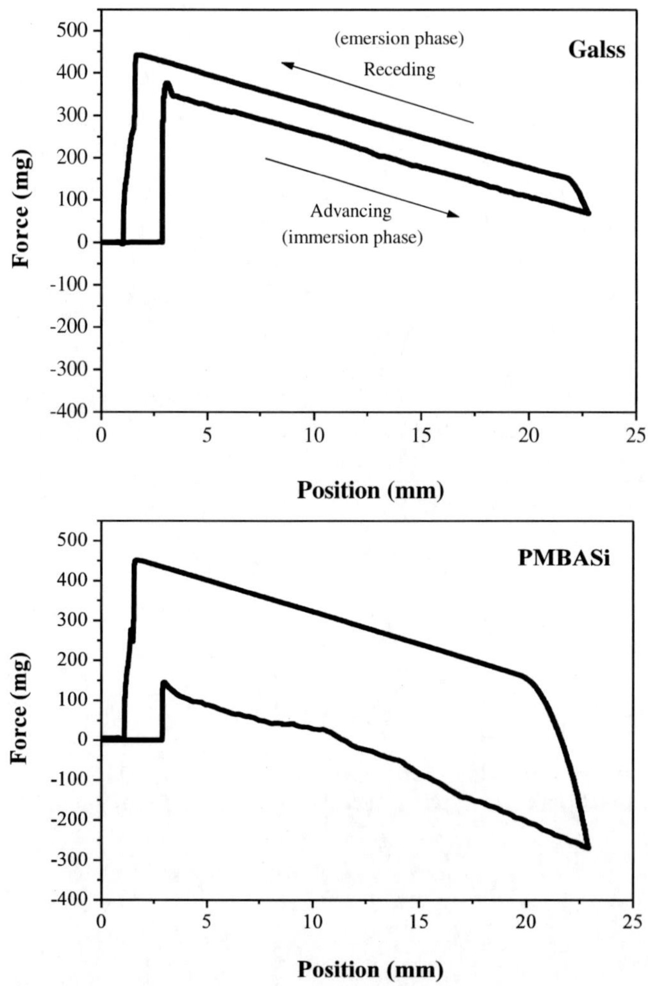

Figure 9. (Continued).

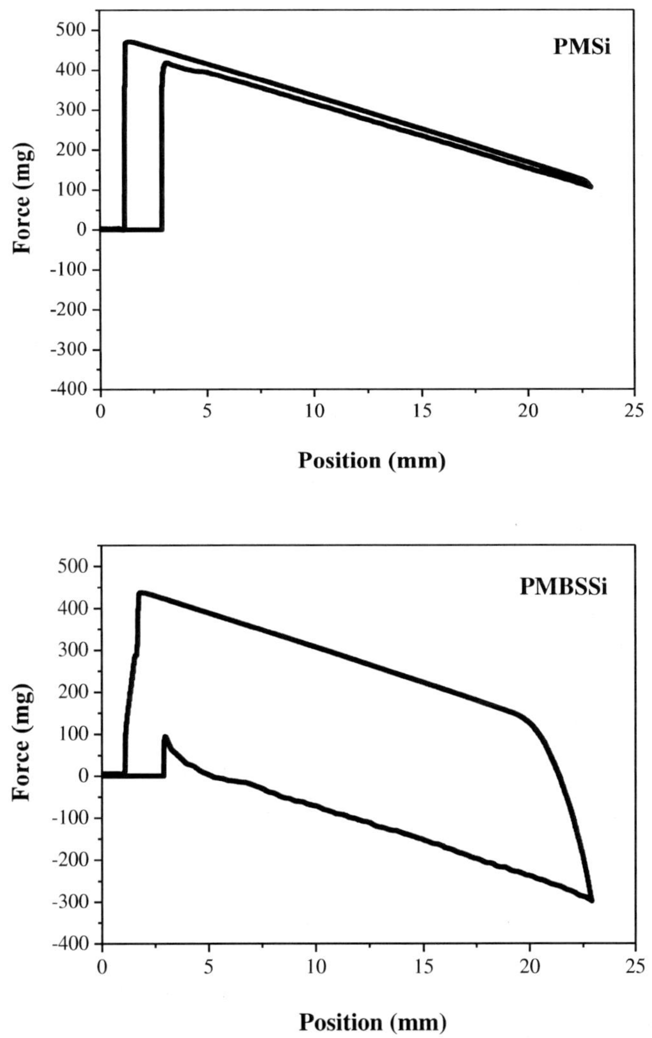

Figure 9. DCA profiles of MPC polymer coated and uncoated glass substrate surfaces being lowered into ultrapure water at a speed of 80 μm s^{-1}. Polymer concentration for coatings was 3.00 mg mL^{-1}, i.e., *high-density coatings*. (Image reproduced, with permission (Elsevier), from ref. 61).

5. Suppression of Protein Adsorption and Cell Adhesion on MPC Polymer Biocompatible Interfaces

5.1 Evaluation of Non-Specific Protein Adsorption

5.1.1. Suppressing Non-Specific Protein Adsorption on the Substrates

The sample surfaces were exposed to a group of proteins to investigate the effects of the surface charge and chemistry on protein adsorption. Based on the electrical nature, we chose pepsin (pI = 1.0) and bovine serum albumin (BSA, pI = 4.7) as typical anionic proteins and ribonuclease A (RNase A, pI = 9.5) and lysozyme (LYZ, pI = 11.0) as typical cationic proteins, respectively. We chose a 1X PBS buffer with a medium ionic strength (about 137 mM) and pH (about 7.1), allowing a better simulation of the typical physiological condition used in various biological applications. Figure 11 shows the adsorbed amounts of four proteins on uncoated, PMBASi-coated, PMSi-coated, and PMBSSi-coated glass substrates, determined with a microBCA protocol which is a widely used method to measure trace amount of protein in solution.

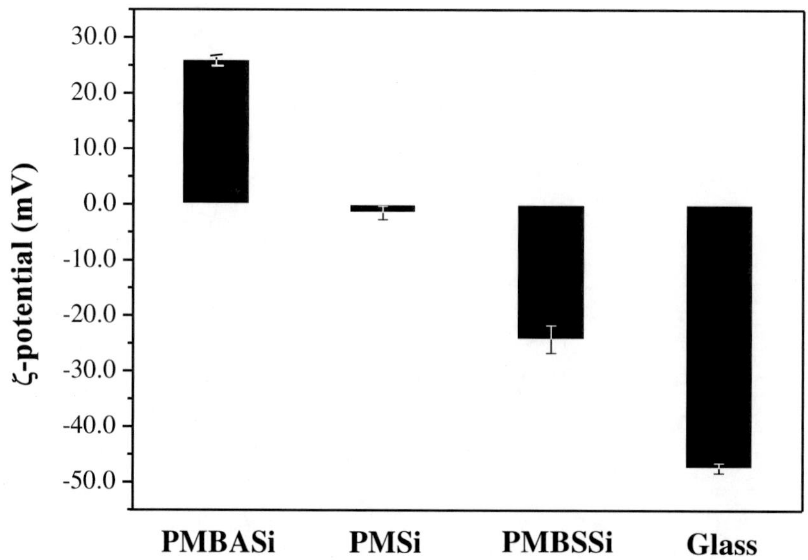

Figure 10. Surface ζ-potentials of glass substrates coated with charged and non-charged MPC polymers in comparison with that of bare glass in aqueous environment. Polymer concentration for coatings was 3.00 mg mL^{-1}, i.e., *high-density coatings*. Data are mean ± SD, n =6. (Image reproduced, with permission (Elsevier), from ref. 61).

The surface coatings remarkably affected the adsorption of the proteins. All proteins, despite their different charge properties, seriously adsorbed to the uncoated glass substrate, exhibiting high adsorbed amounts. In contrast, the adsorbed protein amounts on three MPC copolymers coated surfaces were pronouncedly low in general. After coating, 82—98% reductions in protein adsorption were obtained in comparison with the corresponding behavior on the uncoated glass substrates. Generally, a surface with a ζ-potential of about 26.1 mV (ζ-potential of PMBASi) or −24.2 mV (ζ-potential of PMBSSi) can induce protein adsorption significantly. For example, Ikada et al. reported the amounts of BSA adsorption on the different charged monomer grafted polyethylene (PE) film surfaces [130]. They carried out the protein adsorption experiments under a similar condition as we used in our experiments. Their results indicated that, the adsorbed amount of BSA on the acrylic acid (AA) grafted PE surface with a ζ-potential of −27.3 mV was 0.78 μg cm^{-2}, an amount which is much higher than that on the PMBSSi surface with a similar ζ-potential; and the adsorbed amount of BSA on the *N, N*-dimethylaminopropyl acrylamide (DMAPAA) grafted PE surface with a ζ-potential of 17.0 mV was 4.76 μg cm^{-2}, an amount which is significantly higher than that on the PMBASi surface with a higher ζ-potential of 26.1 mV. This comparison clearly reveals that despite the charged moieties involved, the MPC moieties on the copolymer still played their role effectively in suppression of protein adsorption. In addition, the adsorbed protein amounts on both the cationic PMBASi surface and the anionic PMBSSi surface were in a similar low level as those on the neutral PMSi surface, with no statistical difference significance.

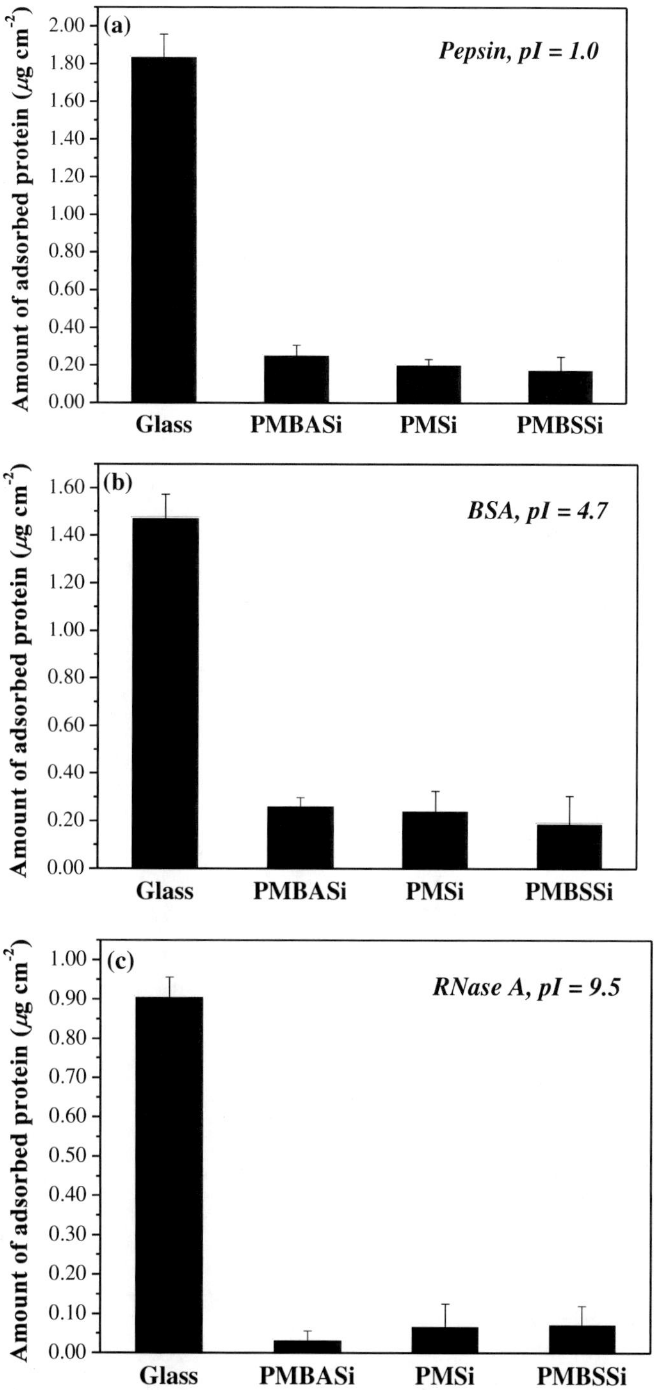

Figure 11. (Continued).

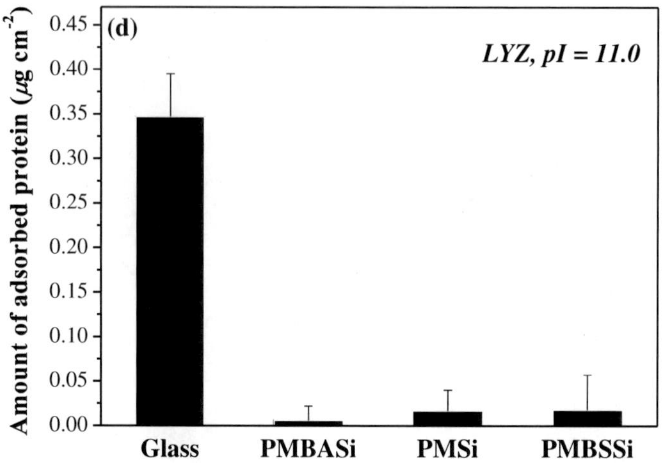

Figure 11. Amounts of typical cationic proteins and anionic proteins adsorbed on glass substrates coated with charged and non-charged MPC polymers in comparison with that of bare glass. (a) pepsin, pI = 1.0; (b) BSA, pI = 4.7; (c) RNase A, pI = 9.5; (d) LYZ, pI = 11.0. Polymer concentration for coatings was 3.00 mg mL^{-1}, i.e., *high-density coatings*. (Image reproduced, with permission (Elsevier), from ref. 61).

AFM observations further revealed the protein adsorption on the surface. AFM images of Figure 12 show the topographies observed in dry and PBS wet conditions of coated and uncoated glass plates after treated with 0.45 g L^{-1} BSA solution. Obviously, after protein treatment for 1 hour at 37 °C, the uncoated glass substrate exhibits a rugged surface in dry condition (Figure 12a), while PMBSSi-coated glass substrate, however, remained relatively flat surfaces at the same scale (Figure 12b, c). The serious protein adsorption on the uncoated glass surface become more obvious after observed in the PBS wet condition (Figure 12d), since the conformation of the adsorbed protein became unfolded and extended after contacting the buffer solution. In contrast, the substrate coated with PMBSSi still kept relatively homogenous surfaces in PBS (Figure 12e, f), due to almost no protein adsorptions. Therefore, BSA is easily absorbed to the uncoated glass surface, but hardly absorbed to the PMBSSi-coated glass surface.

These results indicate that the incorporation of charged moieties would not significantly influence the protein resistant property of a MPC copolymer. We have made a hypothesis that the function of MPC moieties in the suppression of protein adsorption, rather than the electrostatic character resulting from the charged moieties, dominated the interactions between the protein and the charged MPC copolymer surfaces, which will discussed later in details. The results test, support, and expand the hypothesis.

Considering that the MPC mole fractions in both PMBASi and PMBSSi were only about half of that in PMSi, the fact of no big difference in the protein adsorption on the neutral PMSi surface in comparison with on the charged PMBASi or PMBSSi surfaces implies that, an about 50% mole fraction of the MPC moiety in a copolymer is sufficient to construct a protein adsorption resistant surface and additional more fraction seems not to bring more reduction in the protein adsorption. In addition, as the charged PMBASi and PMBSSi contain hydrophobic moieties (i.e. BMA) but the neutral PMSi dose not, the similar level of protein adsorption meanwhile suggests that the introduction of the hydrophobic moieties in a MPC copolymer surface may not result in an increase in the protein adsorption if the MPC moiety is predominant in the copolymer composition. This should be associated with the fact that both the PMBASi and the PMBSSi surfaces had high mobility as elucidated by the DCA analysis (Figure 9 and Table 3). The high mobility facilitates the surface reorientation when the surface is in contact with an aqueous medium, and the hydrophobic components (for example, BMA in both PMBASi and PMBSSi) are buried in the inner of the polymer layer as the hydrophilic components such as MPC moieties migrate to the outmost surface to minimize the surface energy [127, 128]. The MPC moieties are predominant at the surface and thereby dominate the surface properties, especially such as the interactions with proteins when in contact with a protein solution.

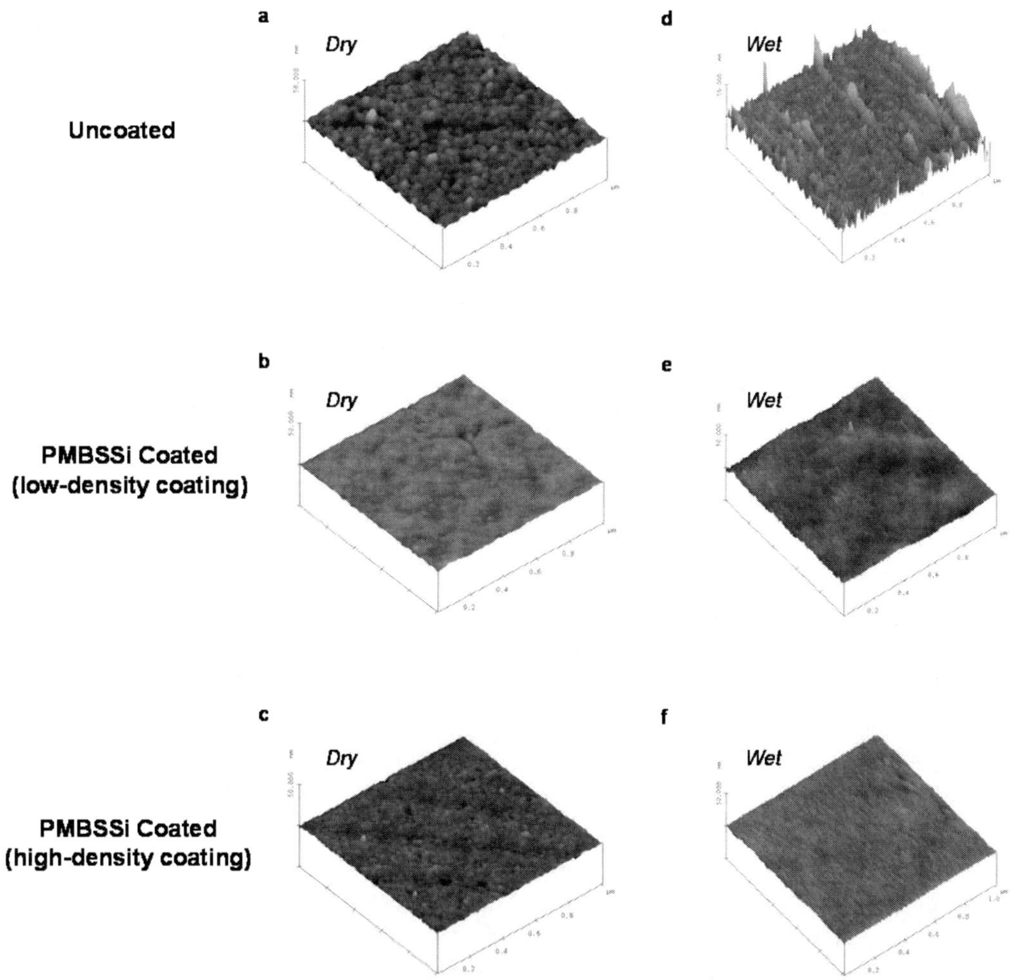

Figure 12. AFM images (1 × 1 μm^2 area) under an ambient (dry) condition and a PBS (wet) condition of (a, d) uncoated glass substrates, (b, e) 0.30 mg mL^{-1} PMBSSi-coated glass substrate (*low-density coating*), and (c, f) 3.00 mg mL^{-1} PMBSSi-coated glass substrate (*high-density coating*), after treated with 0.32 g L^{-1} BSA solution for 1 hour at 37 °C. The z-scale is from 0 to 50 nm.

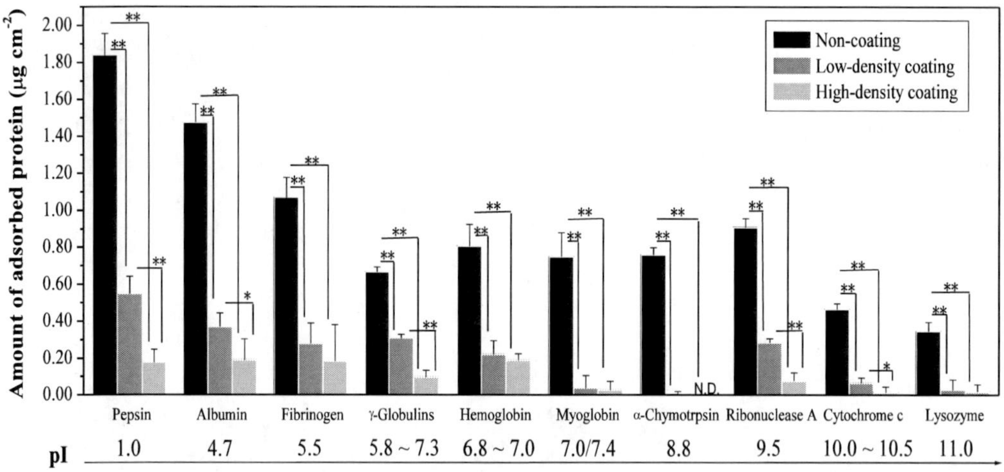

Figure 13. Amounts of various proteins with *pI* values varying from 1.0 to 11.0 absorbed on PMBSSi coated glass substrates with different coating density: *Non-coating* (density = 0.00 μg cm^{-2}), *Low-density coating* (density = 1.19 μg cm^{-2}) and *High-density coating* (density = 11.14 μg cm^{-2}). Data are mean ± SD, n = 9. Statistic difference significance: *, $p < 0.01$; **, $p < 0.001$. (Image reproduced, with permission (ACS), from ref. 60).

Although the adsorbed protein amounts on three coated surfaces had no statistical difference significance with each other, quantitatively they still present a subtle but systematic dependence on the surface charge. When ranking the amounts of protein adsorbed on the different coated surfaces, in the cases of negatively charged pepsin and BSA (Figure 11a, b), the order is PMBASi (cationic) > PMSi (neutral) > PMBSSi (anionic), while in the cases of positively charged RNase A and LYZ (Figure 11c, d), the order is PMBASi (cationic) < PMSi (neutral) < PMBSSi (anionic). This can be correlated to the possible electrostatic interactions between the charged proteins and the charged surfaces [123]. The possible effect from electrostatic interactions would be too small and too limited to influence the overall situation.

5.1.2. Effect of Surface Density

The protein adsorption to a surface depends on the surface properties to some extent. As described previously, the differences in surface properties of PMBSSi coating can be essentially attributed to the difference in surface coating density. Hence, it is useful to obtain a relationship between the coating density and the amount of protein adsorbed on the charged surface. Comparing the data of the amounts of adsorbed protein (Figure 13), we found that after coating with PMBSSi, both the high-density coating and the low-density coating caused a dramatic decrease in the protein adsorption. There was, however, some slight difference in the amounts of protein adsorption between the two cases. All the protein adsorption data indicated that the surface with the high-density coating exhibited much lower amounts of protein adsorption than the surface with the low-density coating. For example, taking pepsin (*pI* = 1.0, anionic) and ribonuclease A (*pI* = 9.5, cationic) as instances, in the case of pepsin, the amounts adsorbed on the low- and high-density coatings were 0.55 μg cm^{-2} and 0.18 μg cm^{-2}, respectively, and in the case of ribonuclease A, they were 0.29 μg cm^{-2} and 0.08 μg cm^{-2}, respectively. That is, the amounts of adsorbed protein in both cases deceased with increasing the PMBSSi coating density. The same tendencies in the effect of the coating density were obtained in all the cases of the other proteins. As shown in Table 4, an increase in the PMBSSi density resulted in an increase in the absolute value of the surface ζ-potential. In some conventional surfaces, as the absolute value of the surface ζ-potential increases, the electrostatic interactions between the surface and the protein would increase, and therefore the risk of protein adsorption would also increase [123]. However, in the PMBSSi coated surfaces, the increase in the surface ζ-potential resulted from the increase in the PMBSSi density, did not lead to more amounts of protein adsorbed to the surfaces, and contrarily, made more reduction in the amount of protein adsorption. This suggests that the PMBSSi density should be the key factor which affects the protein adsorption/resistance on the PMBSSi coated surface. It is known that the PC groups of MPC units play a key role in the suppression of protein adsorption in the MPC polymer coating. More PC groups mean a higher capability to suppress the protein adsorption to some extent. Comparing with the low-density coating, the high-density coating made more PMBSSi react with the glass surface, and thereby afforded a higher composition of MPC units (or PC groups) in the coating layer. This contributed to more reduction in protein adsorption in the high-density coating. In addition, comparing with the high-density coating, the low-density coating has a looser steric structure of coating layer, which may create more chances for some proteins to penetrate into the inner layer of the coating, and consequently lead to an increase in the amount of protein adsorption. Therefore, in the PMBSSi coating system, we consider that the amount of proteins in the PMBSSi surface might depend more on the PMBSSi density than on the surface charge of the coating.

5.1.3. Effect of Buffer Ionic Strength

The ionic strength of a buffer is generally considered to remarkably affect a protein adsorbing from buffer to a charged surface, because counterion concentration affects the degree to which the surface charge is shielded. In order to further elucidate the role of surface charge in the protein adsorption to PMBSSi coated surfaces, the adsorption behaviors were studied in buffers with different ionic strengths. The increase of the ionic strength was obtained by addition of NaCl to 1X D-PBS. Both BSA and LYZ adsorptions were studied as functions of the addition of NaCl. As shown in Figure 14, the amounts of BSA and LYZ adsorbed on the uncoated glass surfaces (non-coating) decreased dramatically with the addition of NaCl in the range of 0 mM to 150 mM. This can be explained by the screened electrostatic interactions occurring at high ionic strength. This reveals that the uncoated glass surface was intensely sensitive to the electrostatic conditions. However, in contrast to the adsorption behavior on the uncoated glass surface, which displayed strong ionic strength sensitivity, the BSA adsorption on both the low- and the high-density coated glass surfaces were not significantly affected by the increase in the ionic strength (Figure 14). The similar behavior also exhibited in the LYZ adsorption, which is shown in Figure 14. These profiles indicate that the PMBSSi coating has an impact on the surface, not only at the physiologic ionic strength but also over a range of the ionic strength. Moreover, the buffer ionic strength has no pronounced influence on protein adsorption to the PMBSSi coated surface, suggesting that this process was not dominated by electrostatic interactions between protein molecules and the surface.

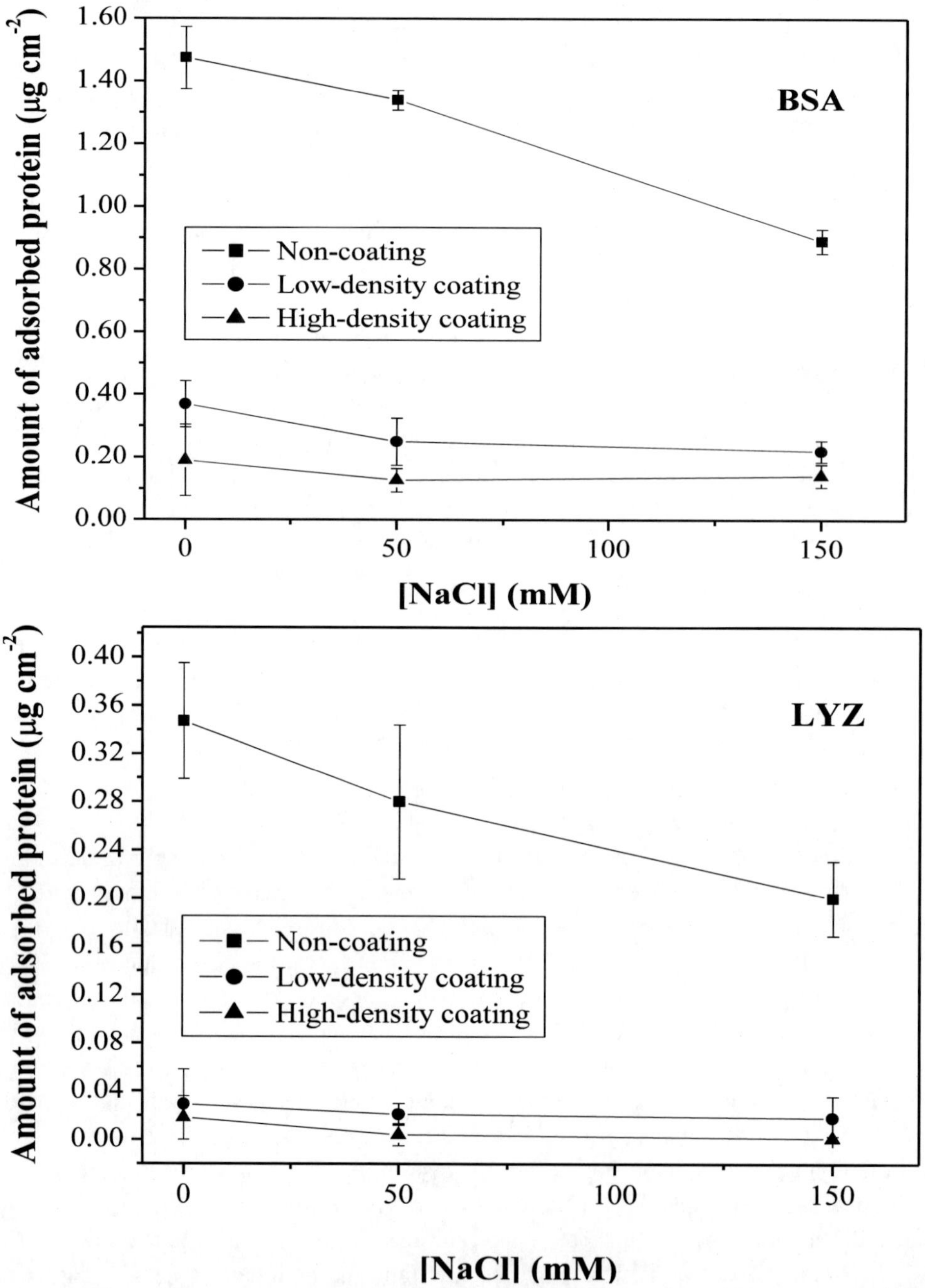

Figure 14. Influence of the buffer ionic strength on the amounts of bovine serum albumin (BSA, pI = 4.7, anionic) and lysozyme (LYZ, pI = 11.0, cationic) adsorbed on different PMBSSi coated glass surfaces (*Non-coating* (density = 0.00 μg cm^{-2}), *Low-density coating* (density = 1.19 μg cm^{-2}) and *High-density coating* (density = 11.14 μg cm^{-2})). The increase of the buffer ionic strength was obtained by addition of NaCl to 1X D-PBS (pH 7.1). [NaCl] represents the added NaCl concentration and the NaCl concentration of original 1X D-PBS (about 137 mM) is not included. Data are mean ± SD, n = 9. (Image reproduced, with permission (ACS), from ref. 60).

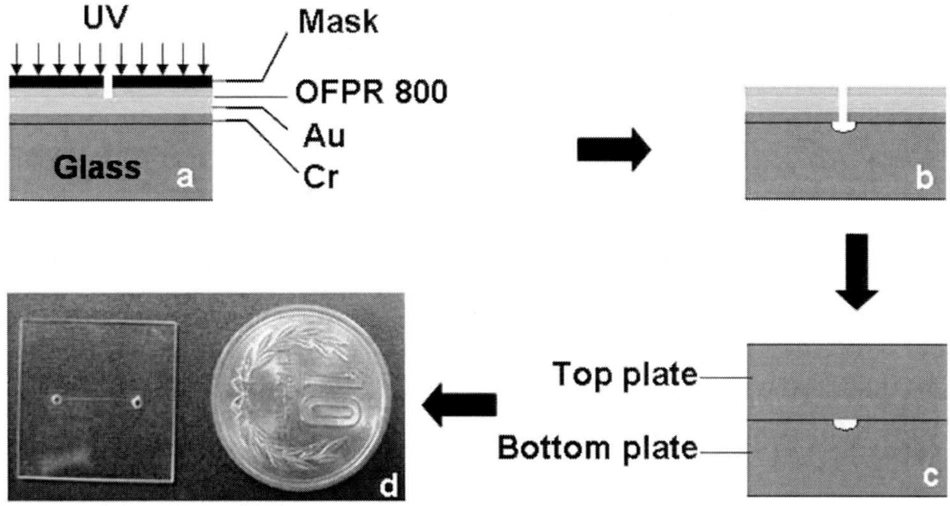

Figure 15. Fabrication of a microchannel on glass substrates ($20 \times 20 \times 0.5$, mm^3). (a) Glass substrate (bottom plate) was coated with Cr, Au, and photoresist (OFPR 800) layers. The channel pattern was transferred by irradiating UV light through a photomask. (b) The microchannel pattern was etched by HF solution after development of the resist, Au, and Cr layers. (c) After removing the resist, Au, and Cr layers, the fabricated bottom plate was fusion bonded with the top plate having reservoirs of inlet and outlet. (d) Picture of the fabricated microchannel.

Although the understanding of the mechanism and the physics of protein resistance specific to MPC-based surface remains in complete in the present time, several hypotheses or theories have been proposed based on these common understandings and our group's research [131-133]. These researches let us be aware of the importance of exploring water interactions among the surface and the around environment to understand the protein resistance. If the water state at the surface is similar to an aqueous solution, protein does not need to release bound water molecules even if protein molecules contact the surface. This means that the hydrophobic interaction does not occur between proteins and the polymer surface. Moreover, the conformational change in the protein's three dimensional structure during protein adsorption on or contact with the surface does not happen. Our research elucidated that MPC-based polymers usually carry higher level of free water fraction on the polymer surface, as is well known that PC groups are highly hydrated. Thus, proteins can contact the surface reversibly, without significant conformational change. The results of the simulation work suggested that PC head groups are very quickly surrounded and polarized by the water molecules when contacting with water [134]. Further, Kitano *et al.* found that the structure of the aqueous solution surrounded PC groups is very close to that of pure water by the Raman spectrum [135].

As charged units (MTAC or PMPS) involved, however, PMBASi and PMBSSi are more complicated than other non-charged MPC polymers such as PMSi. In this work, the experimental results suggest that the amount of surface charge was shown to be a poor determinant of protein adsorption (Figure 11, 12, 13). Accordingly, the effect of electrostatic interactions on the adsorption of the proteins studied is minimal or neglectable (Figure 14). This tests and supports the hypothesis proposed by Lewis A. L. that, even though the surface charge of the MPC polymer coating is changed by incorporation of the monomer having charged groups, if a significant proportion of PC in the composition exists, the interaction of the protein with the charged surface may still be similar to that with a surface of a non-charged MPC polymer coating [136]. This group demonstrated adding cationic charge to MPC polymer with the cationically charged monomer choline methacrylate (CMA), that all polymer coatings containing 0—23% CMA (mole fraction, converted from the weight percentage originally described by the authors in the reference) exhibited small amounts of protein adsorption, with no statistical difference [136]. In the PMBSSi, the mole fraction of MPC monomers is 46%, which is predominant among all components. In contrast, the anionically charged monomer PMPS is only 9% in mole fraction, an amount which is considered incapable of inducing significant amounts of protein adsorption. In addition, in the case of the PMBSSi coating, the increase of the buffer ionic strength by addition of NaCl does not remarkably change the property of the PC dominated interface, so it does not significantly affect the resistance of protein adsorption on the surface. Therefore, the charge in the PMBSSi not only satisfies the requirement for generation of the electrokinetic phenomena but also has no adverse effect on suppression of the protein adsorption.

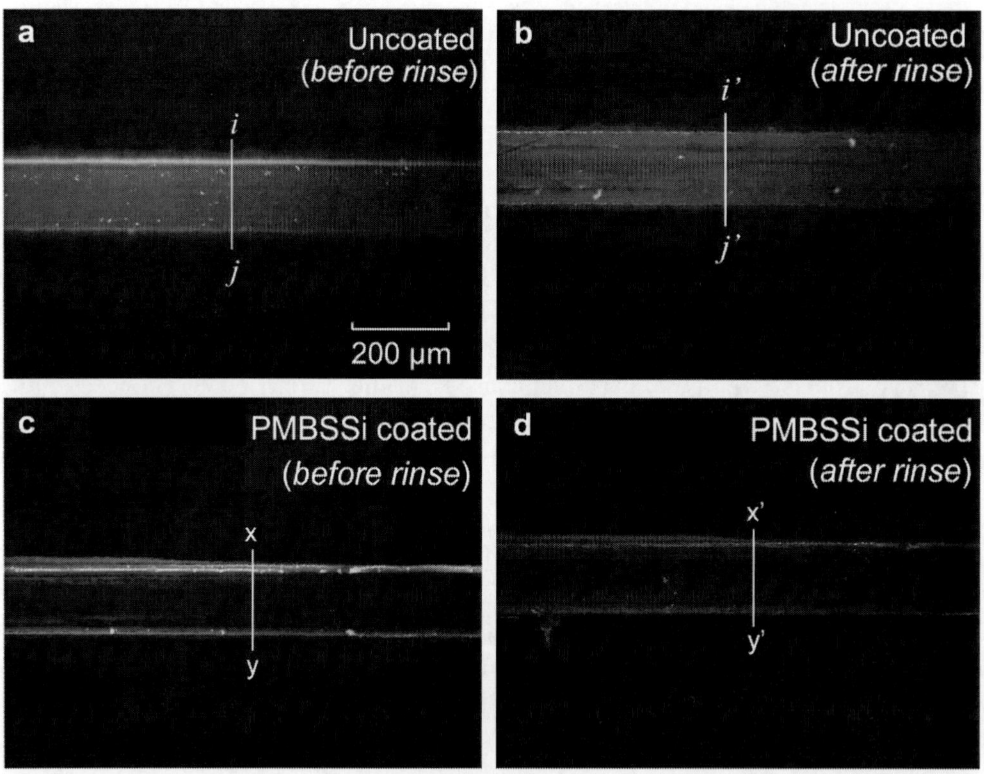

Figure 16. Fluorescence images of protein adsorption in glass microchannels uncoated (the upper, a and b) or coated (the lower, c and d) with PMBSSi (0.30 mg mL^{-1}, *low-density coating*). Firstly, the microchannels were completely filled with 0.45 g L^{-1} FITC-BSA solution, and then incubated for 1h at 4 °C. After venting the FITC-BSA solution (a and c, respectively), the microchannels were rinsed completely four times with 1.2 mL PBS (pH 7.1), (b and d, respectively). (Image partially adopted, with permission (RSC), from ref. 59).

5.1.4. Suppressing Protein Adsorption in Microfluidic Devices

Microchannels were used to qualitatively evaluation of protein adsorption on the MPC polymer interfaces in microfluidic devices. The microchannel was fabricated on the glass substrates by a photolithographic wet etching technique (Figure 15). The MPC polymer interfaces were prepared in the microchannels through a filling coating process described as follows. The microchannel was first cleaned with ethanol which was introduced into the microchannel through the inlet. Then, the MPC polymer solution with the desired concentration was continuously passed through the microchannel. Next, 0.5 mL of the polymer solution was used for each microchannel; finally, a completely filled channel was created. The fine modified channel was realized after the solution in the microchannel underwent complete natural volatilization in air.

A fluorescein conjugated protein—FITC-BSA—was used to visually evaluate the protein adsorption in the microchannel. As a representative, only the results of PMBSSi coatings are shown and discussed here. As demonstrated in Figure 16, in the case of the microchannel without PMBSSi coating, the FITC-BSA was strongly adsorbed in the inner surface of the microchannel, yielding a bright and homogeneous fluorescence emission even after four complete rinses with using 1.2 mL of PBS (Figure 16a, b, and corresponding profiles in Figure 17a, b, respectively). In contrast, from both the observations of fluorescence images (Figure 16c, d, respectively) and the analyses of fluorescence intensity profiles (corresponding profiles in Figure 17c, d, respectively), the microchannel with a really low concentration (0.30 mg mL^{-1}) of PMBSSi coating, showed very low fluorescence emission after venting the protein solution and almost no fluorescence emission was observed after four rinses in the most area of microchannel. Such significant difference in the intensity of fluorescence should be attributed to the PMBSSi coating on the microchannel, although it is difficult to directly confirm the coating on elemental level in microchannel by XPS. In addition, the fluorescence observed in the joint area of chip bonding is possibly due to the relative inhomogeneous surface there. The fluorescence in joint area decreased significantly after rinsing (Figure 16) suggests that most proteins did not firmly adsorb there as the proteins absorbed to uncoated microchannel did. All these results reveal that the PMBSSi coating has good property of suppressing the protein (BSA) adsorption in microchannel.

Similar fluorescence images (data not shown) were obtained in the other MPC polymer (PMSi and PMBASi) coated glass microchannels when treated with the FITC-BSAsolution, which reveals that all three MPC polymer coatings have ability to suppress the protein adsorption in a microchannel as on a coated substrate surface.

5.2. Evaluation of Cell Adhesion

At the molecular level, it is generally believed that the rapid adsorption of proteins to the surface is the initial stage of cell adhesion [117]. In view of the high capability of the three MPC copolymers to suppress protein adsorption, we supposed that they would have corresponding capability to resist the cell adhesion. Mouse fibroblast L929 cells, which are typical adherent cells and are widely used in various biological studies and evaluations as model cells, were employed to evaluate the cell adhesion on the sample surfaces. Figure 18 shows the microscopic images of sample surfaces on which L929 cells were cultured for one day and four days. In the case of uncoated surface, L929 cells gradually adhered, spread, and flattened on the surface after seeded, and then formed confluence after cultured for one day (Figure 18), and finally densely covered the surface four days later (data not shown). In contrast, no cell adhesion was observed on PMBASi, PMSi, and PMBSSi surfaces after cultured for one day and even for four days (Figure 18). This indicates that like the neutral PMSi, both the cationic PMBASi and the anionic PMBSSi possess the high capability to suppress cell adhesion due to the MPC moieties at the surface. Ikada et al. reported that cells could adhere and proliferate on the charged monomer grafted PE surfaces with a ζ-potential larger than 12 mV [130]. Before we performed the evaluation, we had previously speculated that the cationic PMBASi surface might induce cell adhesion because of the possibly increased electrostatic attraction between the positive charged surface of the coating and the negatively charged surface of the cell. Evidently, the evaluation result overturned the speculation, and it seems that the charged moieties in PMBASi and PMBSSi would not cause the effect in inducing cell adhesion. This further supports our aforementioned hypothesis that the electrostatic character resulting from the charged moieties of the copolymers would not dominate the surface behaviors of bioadsorptions.

6. Control of Microfluidic Flow on MPC Polymer Biocompatible Interfaces

6.1. Electroosmotic Flow (EOF) in Microfluidic Devices

Microfluidic systems rely on the high-precision control of microfluidic flow.[115, 137-140] A pressure-driven flow (or hydraulic flow) generated by an external pump is difficult to control since the pressure required depends on the channel radius, and flow velocity differences in the axial easily result in band broadening due to the so-called Taylor dispersion. The electroosmotic flow (EOF), as a state-of-the-art approach widely applied in miniaturized systems for fluid delivery, offers many advantageous features over the pressure-driven pumps: the voltage required is independent of the channel radius; the velocity of the plug-like flow is nearly constant, which makes it easy to avoid the turbulent mixing of fluids; and external pumps or special device structures such as microvalves are not required, which makes miniaturization relatively simpler. The EOF can be induced by applying an external electric field parallel to a surface with a net surface charge density to drive the bulk fluid. For example, in a silica microchannel filled with a solution whose pH is greater than the pK_a value of silica (2.8), the channel wall will acquire a negative surface charge due to the dissociation of surface silanol groups and will leave behind a layer of positive ions known as the electrical double layer (EDL). The EDL contains an excess of mobile positive ions and the potential difference across this mobile part of the EDL is known as the zeta potential (ζ). When an axial electrical field (E) is applied, these positive ions migrate towards the cathode; they are then transmitted to the bulk of the fluid through the effect of viscosity, that is, the EOF is generated. The velocity of the EOF, u_{EOF}, can be given by the Helmholtz-Smoluchowski equation:

$$u_{EOF} = \mu_{EOF} E = \frac{\zeta \varepsilon}{\eta} E$$

where μ_{EOF} is the EOF mobility; ζ, the zeta potential of the channel wall; and ε and η, the relative permittivity and viscosity of the buffer solution, respectively.

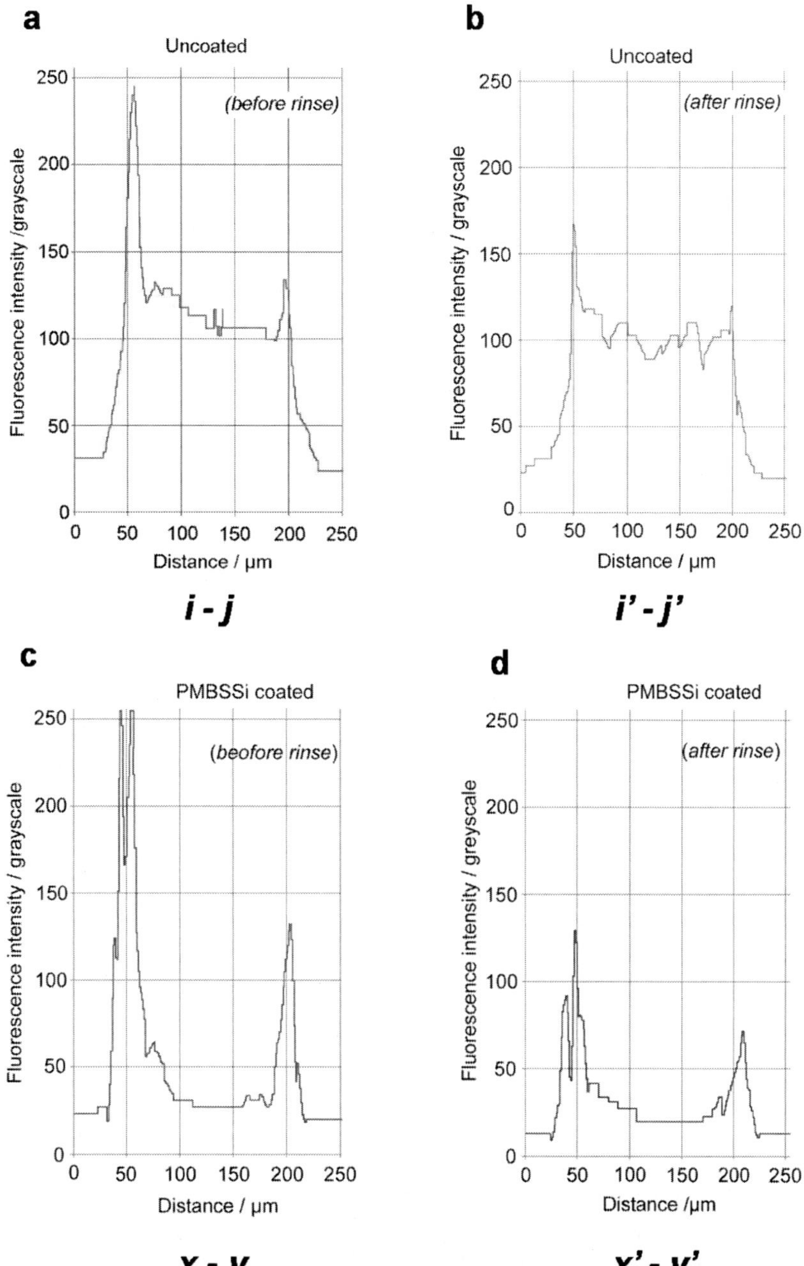

Figure 17. Fluorescence intensity profiles of a) line *i* – *j*, b) line *i'* – *j'*, c) line *x-y*, and c) line *x'-y'* in Figure 16, respectively.

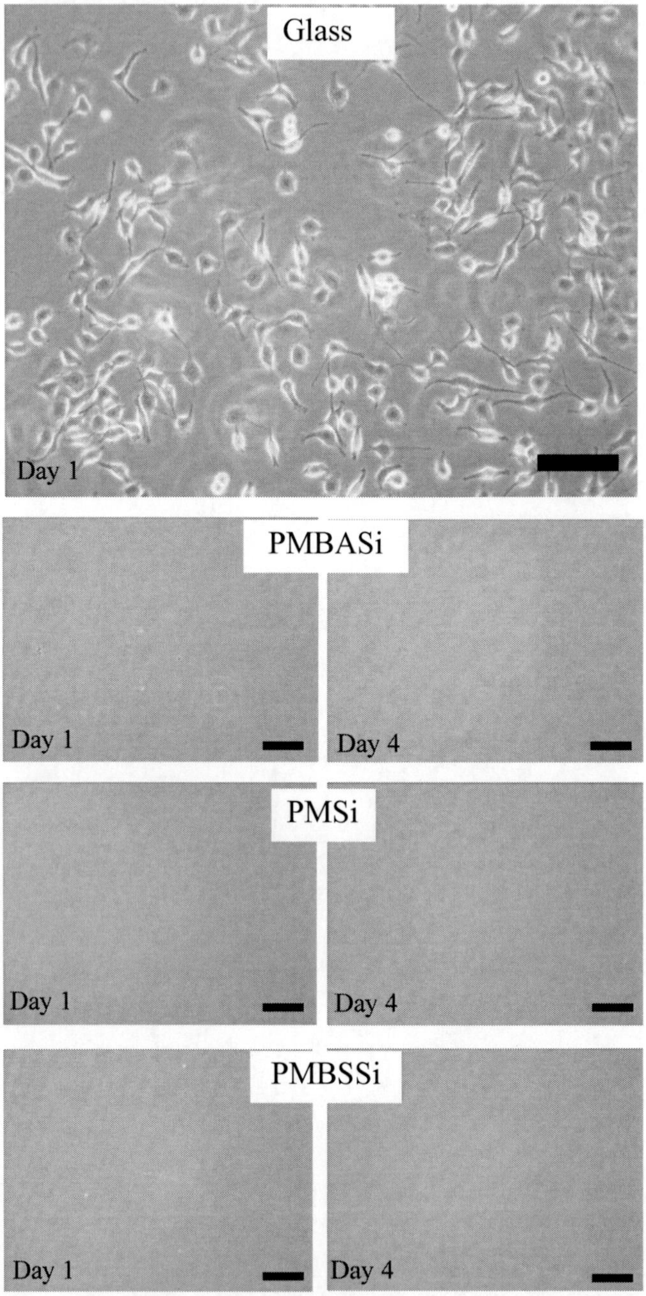

Figure 18. Microscopic images of MPC polymer-coated and -uncoated surfaces on which L929 cells were cultured for 1 day and 4 days. The scale bar represents 100 μm. (Image reproduced, with permission (Elsevier), from ref. 61).

From the abovementioned relationship, the magnitude and direction of the EOF could be controlled by adjusting these parameters individually or together. For example, the EOF could be controlled by modifying the ζ-potential of the channel wall. As a primary strategy, wall coating has been successfully applied to modify ζ and thereby influence the EOF [33]. To retain or reverse EOF, charged polymer coatings have been applied. Most of charged polymeric coatings reported are cationic in nature (such as polyethyleneimine, polyarginine), because the positively charged polymer is easy to strongly adsorb to the silica-based substrates. These coatings are usually utilized to reverse the EOF; while for retaining the EOF, negatively charged coatings may be preferred. To construct negatively charged surfaces, successive multiple ionic layer (SMIL) adsorbed coatings have been used. However, the procedures of these coatings are usually not simple. In addition, these studies were usually aimed at improving the electrophoresis or electrochromatography efficiency rather than the flow control.

As aforementioned, another problem of EOF control in miniaturized systems for biological applications is biomolecule adsorption, which may adversely influence the overall behaviour of the EOF by altering the ζ-potential and causing temporal anomalies that are undesirable invariants in flow control. Therefore, for successfully controlling the EOF, the biomolecule adsorption (especially protein adsorption) inside the device should be reduced. The MPC polymers, PMBASi, PMSi and PMBSSi, have potentials to be applied to control the EOF in microfluidic devices for biological applications, because the interfaces constructed by these MPC polymers are not only charged but also protein-adsorption resistant.

6.2. EOF on MPC Polymer Biocompatible Interfaces

6.2.1. Profile of EOF

As an essential approach, the control of the EOF by modulating the applied voltage was conducted on the single-channel chip (Figure 15) whose inner surface was modified using the MPC polymer (PMBSSi). To monitor the EOF, neutral polystyrene microspheres were used as EOF tracers. The velocities and directions of the plug-flows were modulated smoothly by adjusting the magnitude and direction of the electric voltage. The tracers maintained a straightforward path throughout the measurement as a feature of plug-flow. The EOF control was so sensitive that no inertia was observed when the applied voltage was turned off or the direction of the applied voltage was reversed (Figure 19).

6.2.2. EOF Mobility

To generate an EOF, the surface of the microchannel should have a certain amount of charge density. As aforementioned, the uncoated glass substrate surface is negatively charged with a ζ-potential of –47.4 ± 0.9 mV at neutral pH. Accordingly, the glass microchannel exhibited a considerably large cathodic EOF with a mobility of $(1.9 \pm 0.1) \times 10^{-4}$ cm^2 V^{-1} s^{-1} at pH 7.0. The EOF determination indicated that the PMBSSi-coated glass surface retained a significant amount of cathodic EOF. For example, in the case of PBS, pH 7.0, the EOF mobility of the microchannel with a high-density PMBSSi coating was $(1.0 \pm 0.1) \times 10^{-4}$ cm^2 V^{-1} s^{-1}, which is roughly more than half (53%) that of the uncoated microchannel at the same buffer condition. Besides, this is in good agreement with the result of the ζ-potential. After modification by PMBSSi (3.00 mg mL^{-1}, *high-density coating)*, the ζ-potential was altered to –24.2 ± 2.5 mV (at neutral pH), which is approximately 51% of that of the uncoated glass plate.

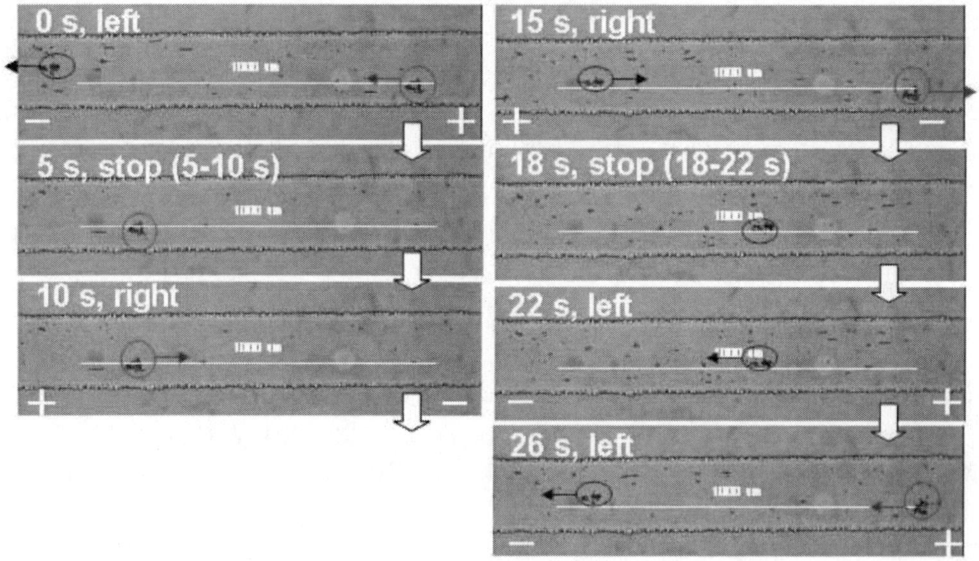

Figure 19. Pictures of the EOF at different time points in the microchannel coated with 0.30 mg mL^{-1} of PMBSSi (low-*density coating*) under an applied voltage of 130 V. The buffer solution is PBS (adjusted to pH 7.0). The polystyrene microspheres with an average diameter of 6.0 μm are used as tracers. The round and elliptical circles indicate the positions of two microsphere groups at a certain distance and the arrowheads denote the direction of the motion. The white line is a scale bar of 1000 μm as well as a path indicator. The applied voltage is turned off for a while; following this, the direction of the applied electric field is changed at approximately every third moment. The plus and minus signs indicate the direction of the applied electric field. (Image reproduced, with permission (RSC), from ref. 59).

6.2.3. Control of EOF

The curves of the ζ-potentials and the EOF mobilities generated on the MPC polymer interfaces with different charge conditions are shown in Figure 20. As mentioned earlier, all the ζ-potentials and the EOF mobility in Figure 20 were determined at neutral pH. It was clearly proved that both the parameters—ζ-potential and EOF mobility—decreased along the following sequence (Figure 20a): 3.00 mg mL^{-1} PMBSSi (*high-density coating*), 0.30 mg mL^{-1} PMBSSi (*low-density coating)*, 0.30 mg mL^{-1} PMSi (*low-density coating)* and 3.00 mg mL^{-1} PMSi (*high-density coating)*, depending on the concentrations of the modification solutions and the charge conditions of the polymers. The tendencies of both the parameters were perfectly consistent with each other. Moreover, the relationship of the ζ-potential and the EOF mobility on the uncoated and coated surfaces demonstrates a liner fit (Figure 20b), which is in accord with the theoretical linear relationship between the ζ-potential and the EOF mobility shown in the Helmholtz-Smoluchowski equation.

The interfaces constructed by PMBSSi and PMSi exhibited a considerably reasonable capability to adjust the ζ-potential and EOF mobility. The microchannels coated by PMBSSi maintained a significant cathodic EOF, while those coated by PMSi exhibited a relatively low cathodic EOF. Based on these results, we attempted to control the EOF with respect to the surface charge conditions, which could be easily controlled by the concentration of modification solutions and polymer designing rather than by other complicated procedures [141-143]. The tendency curves of the ζ-potential and EOF mobility, depending on the concentration of the polymer solution for modification, could provide a feasible scheme to simultaneously generate or control the EOF with distinct velocities in several channels under a single applied voltage.

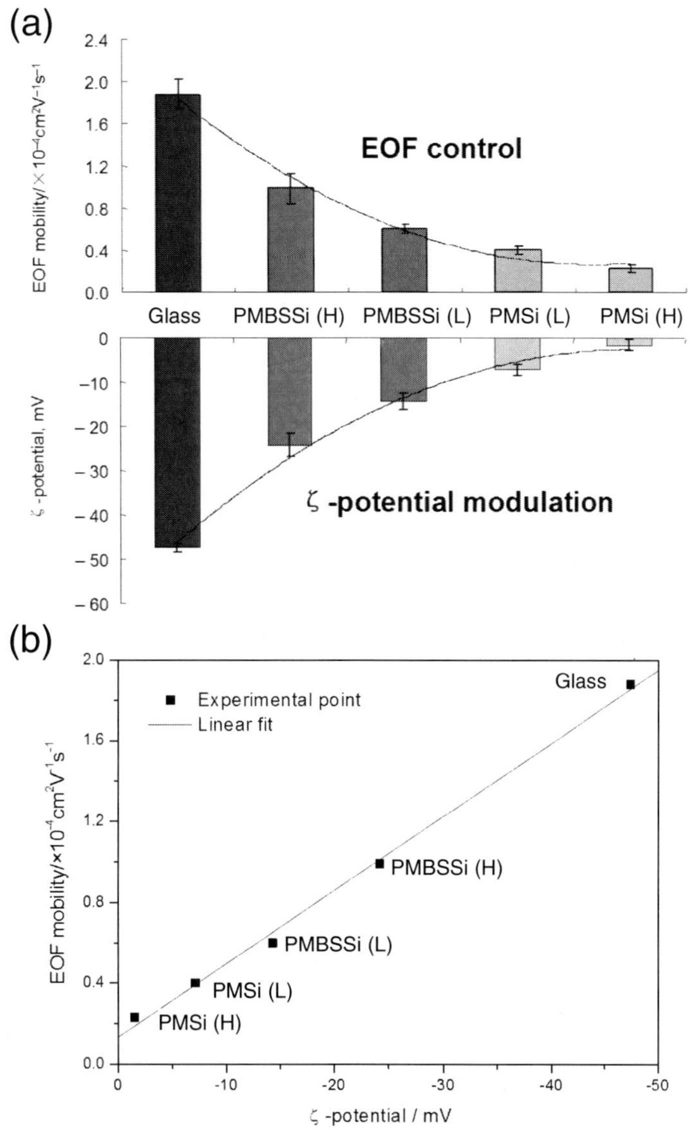

Figure 20. a) ζ-potentials and EOF mobilities of the glass microchannel coated by PMBSSi and PMSi with different coating densities (H: *high-density coating*; L: *low-density coating*) at neutral pH, and b) a linear fit of ζ-potentials and EOF mobilities. (Image partially adopted, with permission (RSC), from ref. 59).

6.2.4. Stability of EOF

The stability during operation and in repetitive runs is another crucial problem of the EOF. However, uncoated silica-based substrates usually show unsatisfactory results with respect to this because of the adsorption of the sample and buffer components at the surface, which alters the ζ-potential and thus the EOF. The effects of certain modifications, especially dynamic coatings, weaken gradually over time and these modifications need to be reapplied.

This usually causes abnormalities in the flow, thereby making it difficult to control. The stabilities of the EOF in the microchannels coated by PMBSSi and PMSi with different concentrations were investigated for four consecutive days with continuous operation for 2 h per day in PBS, pH 7.0. The result shown in Figure 21 indicates a relatively stable EOF during the investigation in all the cases. We believe that this is not only because of the stable chemical bonding between the substrate and the MPTMSi units in MPC polymers but also because of the inertness of such interfaces to buffers.

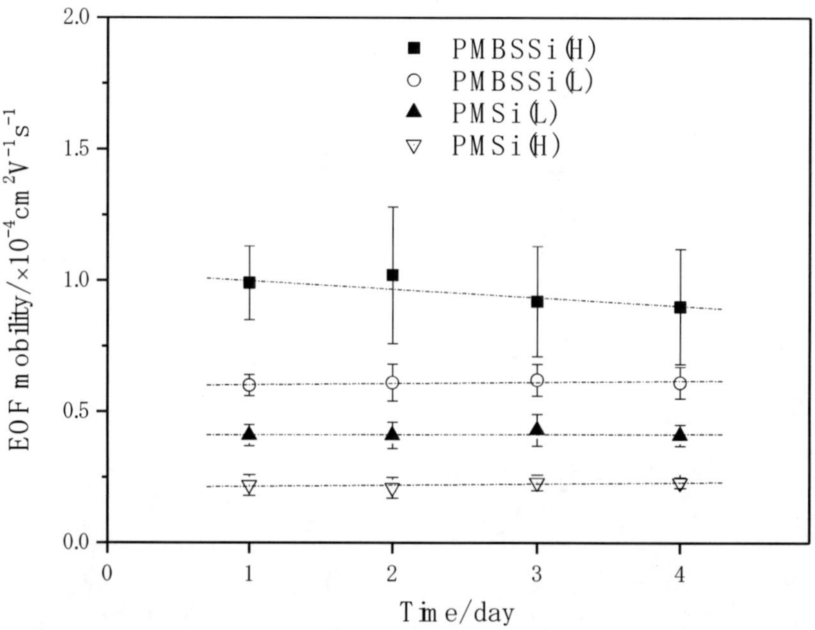

Figure 21. The stability of EOF mobility in the microchannels modified by PMBSSi and PMSi with coating densities (H: *high-density coating*; L: *low-density coating*). EOF mobilities were measured over four consecutive days with continuous operation for 2 h per day. The measurements were performed at room temperature with PBS (pH 7.0). (Image reproduced, with permission (RSC), from ref. 59).

6.2.5. Effect of Buffer pH on EOF

It has been reported that the surface ζ-potential and the EOF mobility of the bare glass surf-ace are significantly influenced by the change in the pH [144, 145]. This can be attributed to the protonation or deprotonation of silanol sites on the glass surface. A similar result was obtained in the case of the uncoated glass in our investigations. As seen in Figure 22, the EOF mobility in the uncoated glass microchannel increased remarkably with the pH in the range of 2 to 7, although it decreased to some extent between pH 8 and 10. However, in contrast to the bare glass microchannel, which displays strong pH sensitivity, the EOF mobilities in both the PMBSSi and PMSi-coated microchannels were only slightly affected by the increase in the pH (Figure 22). This profile indicates that the PMBSSi and PMSi modifications have an impact on the surface not only at neutral pH but also over a range of pH. The modifications are stable and the fluid environment has no pronounced influence on the EOF.

6.2.6. *Effect of Protein Adsorption on EOF*

To investigate the effect of protein adsorption on the EOF, a process for protein treatment was applied in the microchannels with BSA as follows. The BSA solution (PBS as buffer) has a concentration of 0.45 g L^{-1}. For comparison, both the coated and uncoated (as control) microchannels were prepared. These were first rinsed by 2 mL of PBS. 0.5 mL of the BSA solution was delivered to pass through the channel continuously and finally a completely filled microchannel was created. The completely filled microchannel was incubated at 4 °C for 1 h in order to achieve sufficient contact between the protein and the inner surface of the microchannel. After incubation, the BSA solution inside the channel was drained out, followed by consecutive rinses with PBS four times with volumes of 0.1 mL, 0.2 mL, 0.4 mL and 0.5 mL, respectively. Then further EOF measurements were conducted.

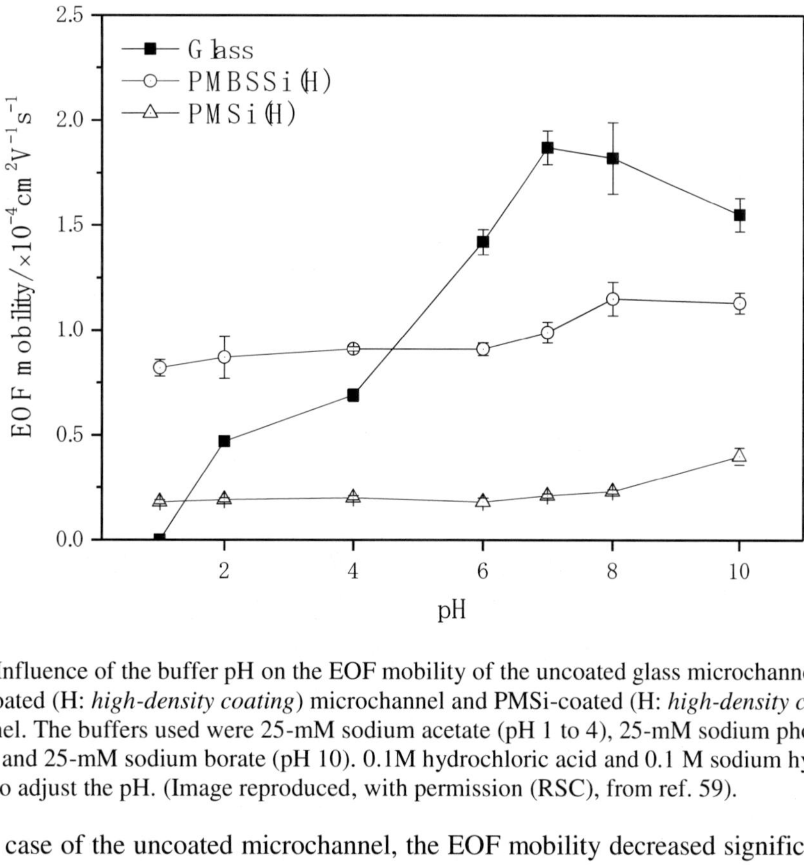

Figure 22. Influence of the buffer pH on the EOF mobility of the uncoated glass microchannel, PMBSSi-coated (H: *high-density coating*) microchannel and PMSi-coated (H: *high-density coating*) microchannel. The buffers used were 25-mM sodium acetate (pH 1 to 4), 25-mM sodium phosphate (pH 6 to 8) and 25-mM sodium borate (pH 10). 0.1M hydrochloric acid and 0.1 M sodium hydroxide were used to adjust the pH. (Image reproduced, with permission (RSC), from ref. 59).

In the case of the uncoated microchannel, the EOF mobility decreased significantly after being treated with a protein solution (Figure 23). This is due to the heavy protein adsorption in the uncoated glass microchannel. However, in the case of the PMBSSi-coated microchannel, the level of the EOF mobility was maintained well and only a very small reduction was displayed post treatment with BSA (Figure 23). This is because the PMBSSi modification effectively suppressed the protein adsorption and the consequent variation in the EOF mobility that resulted from protein adsorption.

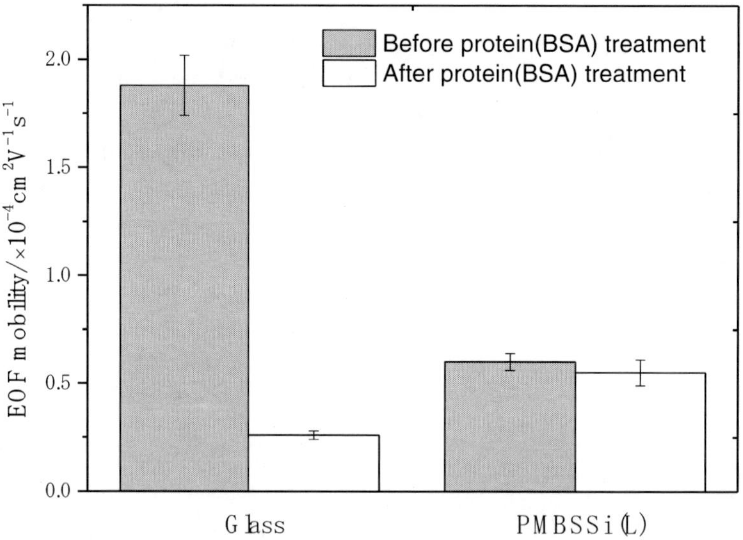

Figure 23. Variations in EOF mobility in uncoated and PMBSSi coated (*low-density coating*) microchannels before and after treatment with BSA solution. First, the EOF mobilities in the microchannels were measured before protein treatment. Then, the microchannels were completely filled with 0.45 g/L of BSA solution and incubated for 1 h at 4 °C. After venting the BSA solution, the microchannels were rinsed with 1.2 mL of PBS (pH 7.1), and then, the EOF mobilities in the microchannels were measured again (after protein treatment). (Image adopted, with permission (RSC), from ref. 59) .

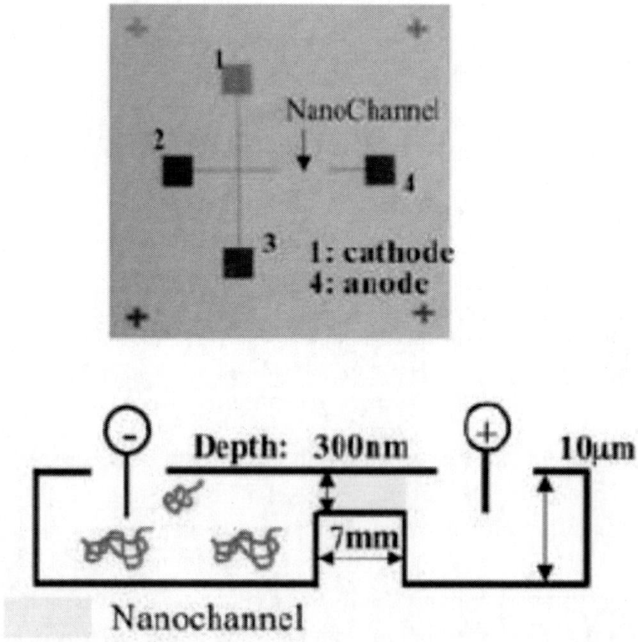

(a)

Figure 24. (Continued).

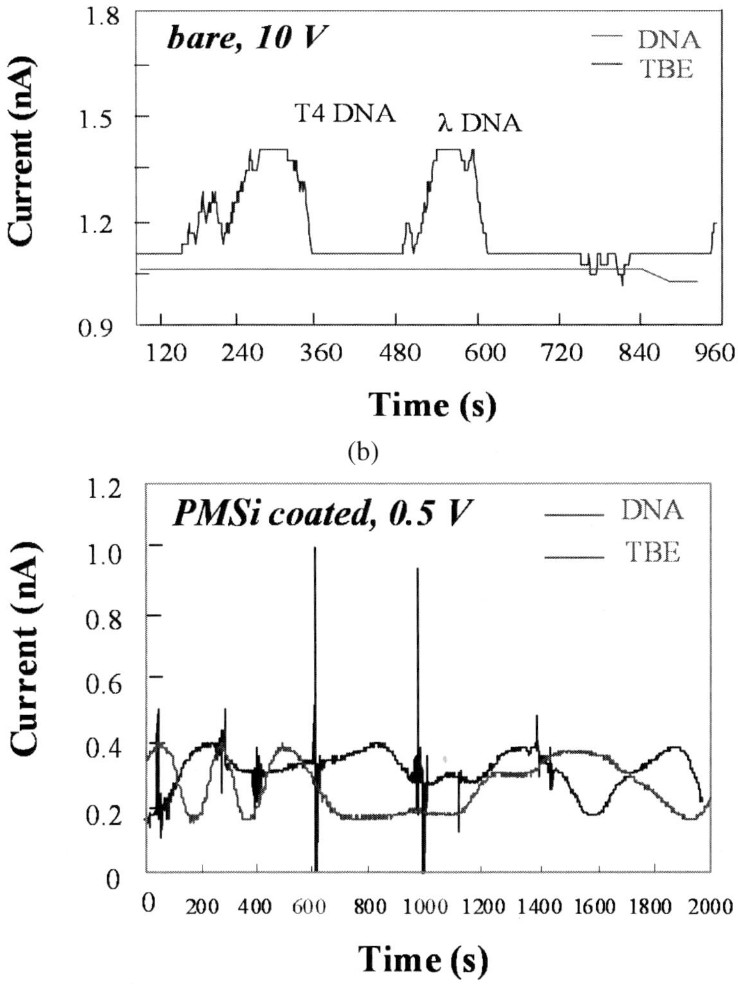

(c)

Figure 24. DNA separation in a nanochannel with MPC polymer (PMSi) interface: (a) photograph of DNA separation chip and schematics of nanochannel design; and DNA separation by electrophoresis on microchips with (b) bare nanochannel at 10 V in comparison with (c) PMSi-coated nanochannel at 0.5 V. (Image partially adopted, with permission (JAACC), from ref. 147) .

7. DNA Detection in Nanochannels with MPC Polymer Biocompatible Interfaces

Microchip electrophoresis is a useful separation tool for genomics and proteomics. Rapid detection of DNA after electrophoretic separation has been achieved using a microchip with a nanochannel having an MPC polymer (PMSi) interface by Shimizu *et al.* [146-148]. The chip was designed for electrical detection, which is a nondestructive detection method because a fluorescence-labeled substance is not necessary. Figure 24a shows a photograph and the channel design of the glass microchip used for DNA separation and detection. The microchannel for electrophoretic separation of DNA was 10 μm in depth and 100 μm in width, and the nanochannel for detection of DNA was 300 nm in depth, 1 μm in width, and 7 mm in length. To construct the MPC polymer interface, 0.5 mg mL^{-1} of PMSi ethanol solution was passed through the channel for 3 min, and then, the chip was dried for 4 h at 70 ºC after venting the polymer solution. 0.4 mg mL^{-1} of T4 DNA (165.6 kbp) and 0.4 mg mL^{-1} λ DNA (48.5 kbp) were prepared in tris-borate-EDTA (TBE) buffer as stock solutions. The DNA separation solution was 0.5x TBE, catalase, glucose oxidase, glucose, and DNA. Voltage was applied between electrodes 1 and 4 of the chip, as shown in Figure 24a, and the current on the anode was measured potentiostatically. T4 DNA and λ DNA were detected based on the increase in the current flowing through the nanochannel. After electrophoretic separation, the separated DNA entered the nanochannel, and the detected current, which was limited by the nanochannel, increased because of the change in the three-dimensional configuration. Broad peaks of T4 DNA and λ DNA were detected in the case of the bare nanochannel (Figure 24b), whereas sharp peaks of T4 DNA and λ DNA were detected in the case of the PMSi-coated nanochannel (Figure 24c). This is because stronger EOF effect decreased the electrophoretic velocity in the bare nanochannel because of a larger ζ-potential (about −47 mV) as compared with the PMSi-coated nanochannel, which had a smaller ζ-potential (about −10 mV), and consequently a weaker EOF effect. Further, the bare chip could not be used after one separation operation, whereas the PMSi-coated chip could be used several times after simply cleaning it ultrasonically, which is made possible by considerably decreased DNA adsorption on the nanochannel by the MPC polymer interface [149].

CONCLUSION

Biocompatible interfaces based on silanization of cationic, neutral, and anionic MPC-MPTMSi polymers on glass-based substrates and microfluidic devices can overcome not only serious problems, such as nonspecific adsorptions, but can also help achieve various functions such as control of microfluidic flow and on-chip biomedical applications such as DNA detection, which would significantly contribute to realize miniaturized and integrated biological and biomedical applications.

ACKNOWLEDGMENTS

This work was partially supported by Funds for the Development of Human Recourse in Science and Technology of Japan Science and Technology Agency (JST). The content of the chapter is partially adapted from our published papers [59-61, 148].

REFERENCES

[1] Whitesides GM. The origins and the future of microfluidics. *Nature* 2006;442:368-73.

[2] Kung CY, Barnes MD, Lermer N, Whitten WB, Ramsey JM. Confinement and manipulation of individual molecules in attoliter volumes. *Anal Chem* 1998;70:658-61.

[3] Jacobson SC, Hergenroder R, Koutny LB, Ramsey JM. High-Speed Separations on a Microchip. *Anal Chem* 1994;66:1114-8.

[4] Yang PD, Rizvi AH, Messer B, Chmelka BF, Whitesides GM, Stucky GD. Patterning porous oxides within microchannel networks. *Adv Mater* 2001;13:427-31.

[5] Bousse L, Mouradian S, Minalla A, Yee H, Williams K, Dubrow R. Protein sizing on a microchip. *Anal Chem* 2001;73:1207-12.

[6] Obeid PJ, Christopoulos TK, Crabtree HJ, Backhouse CJ. Microfabricated device for DNA and RNA amplification by continuous-flow polymerase chain reaction and reverse transcription-polymerase chain reaction with cycle number selection. *Anal Chem* 2003;75:288-95.

[7] Seong GH, Heo J, Crooks RM. Measurement of enzyme kinetics using a continuous-flow microfluidic system. *Anal Chem* 2003;75:3161-7.

[8] Keng PY, Chen SP, Ding HJ, Sadeghi S, Shah GJ, Dooraghi A, et al. Micro-chemical synthesis of molecular probes on an electronic microfluidic device. *P Natl Acad Sci* USA 2012;109:690-5.

[9] Jiang XY, Ng JMK, Stroock AD, Dertinger SKW, Whitesides GM. A miniaturized, parallel, serially diluted immunoassay for analyzing multiple antigens. *J Am Chem Soc* 2003;125:5294-5.

[10] Sia SK, Linder V, Parviz BA, Siegel A, Whitesides GM. An integrated approach to a portable and low-cost immunoassay for resource-poor settings. *Angew Chem Int Edit* 2004;43:498-502.

[11] McClain MA, Culbertson CT, Jacobson SC, Allbritton NL, Sims CE, Ramsey JM. Microfluidic devices for the high-throughput chemical analysis of cells. *Anal Chem* 2003;75:5646-55.

[12] Erickson D, Sinton D, Li DQ. A miniaturized high-voltage integrated power supply for portable microfluidic applications. *Lab Chip* 2004;4:87-90.

[13] Chin CD, Laksanasopin T, Cheung YK, Steinmiller D, Linder V, Parsa H, et al. Microfluidics-based diagnostics of infectious diseases in the developing world. *Nat Med* 2011;17:1015-U138.

[14] Manz A, Harrison DJ, Verpoorte EMJ, Fettinger JC, Paulus A, Ludi H, et al. Planar Chips Technology for Miniaturization and Integration of Separation Techniques into Monitoring Systems - Capillary Electrophoresis on a Chip. *J Chromatogr* 1992;593:253-8.

[15] Harrison DJ, Fluri K, Seiler K, Fan ZH, Effenhauser CS, Manz A. Micromachining a Miniaturized Capillary Electrophoresis-Based Chemical-Analysis System on a Chip. *Science* 1993;261:895-7.

[16] Kopp MU, de Mello AJ, Manz A. Chemical amplification: Continuous-flow PCR on a chip. *Science* 1998;280:1046-8.

[17] Burns MA, Johnson BN, Brahmasandra SN, Handique K, Webster JR, Krishnan M, et al. An integrated nanoliter DNA analysis device. *Science* 1998;282:484-7.

[18] Beebe DJ, Moore JS, Bauer JM, Yu Q, Liu RH, Devadoss C, et al. Functional hydrogel structures for autonomous flow control inside microfluidic channels. *Nature* 2000;404:588-+.

[19] Thorsen T, Maerkl SJ, Quake SR. Microfluidic large-scale integration. *Science* 2002;298:580-4.

[20] MacDonald MP, Spalding GC, Dholakia K. Microfluidic sorting in an optical lattice. *Nature* 2003;426:421-4.

[21] Ottesen EA, Hong JW, Quake SR, Leadbetter JR. Microfluidic digital PCR enables multigene analysis of individual environmental bacteria. *Science* 2006;314:1464-7.

[22] Prakash M, Gershenfeld N. Microfluidic bubble logic. *Science* 2007;315:832-5.

[23] Zhang J, Coulston RJ, Jones ST, Geng J, Scherman OA, Abell C. One-Step Fabrication of Supramolecular Microcapsules from Microfluidic Droplets. *Science* 2012;335:690-4.

[24] Tadmor AD, Ottesen EA, Leadbetter JR, Phillips R. Probing Individual Environmental Bacteria for Viruses by Using Microfluidic Digital PCR. *Science* 2011;333:58-62.

[25] Miller OJ, El Harrak A, Mangeat T, Baret JC, Frenz L, El Debs B, et al. High-resolution dose-response screening using droplet-based microfluidics. *P Natl Acad Sci* USA 2012;109:378-83.

[26] Reyes DR, Iossifidis D, Auroux PA, Manz A. Micro total analysis systems. 1. Introduction, theory, and technology. *Anal Chem* 2002;74:2623-36.

[27] Auroux PA, Iossifidis D, Reyes DR, Manz A. Micro total analysis systems. 2. Analytical standard operations and applications. *Anal Chem* 2002;74:2637-52.

[28] Mukhopadhyay R. When microfluidic devices go bad - How does fouling occur in microfluidic devices, and what can be done about it? *Anal Chem* 2005;77:429a-32a.

[29] Popat KC, Desai TA. Poly(ethylene glycol) interfaces: an approach for enhanced performance of microfluidic systems. *Biosens Bioelectron* 2004;19:1037-44.

[30] Larsericsdotter H, Jansson O, Zhukov A, Areskoug D, Oscarsson S, Buijs J. Optimizing the surface plasmon resonance/mass spectrometry interface for functional proteomics applications: How to avoid and utilize nonspecific adsorption. *Proteomics* 2006;6:2355-64.

[31] Roach LS, Song H, Ismagilov RF. Controlling nonspecific protein adsorption in a plug-based microfluidic system by controlling interfacial chemistry using fluorous-phase surfactants. *Anal Chem* 2005;77:785-96.

[32] Liu JK, Lee ML. Permanent surface modification of polymeric capillary electrophoresis microchips for protein and peptide analysis. *Electrophoresis* 2006;27:3533-46.

[33] Dolnik V. Wall coating for capillary electrophoresis on microchips. *Electrophoresis* 2004;25:3589-601.

[34] Belder D, Ludwig M. Surface modification in microchip electrophoresis. *Electrophoresis* 2003;24:3595-606.

[35] Wu DP, Zhao BX, Dai ZP, Qin JH, Lin BC. Grafting epoxy-modified hydrophilic polymers onto poly(dimethylsiloxane) microfluidic chip to resist nonspecific protein adsorption. *Lab Chip* 2006;6:942-7.

[36] Wong I, Ho CM. Surface molecular property modifications for poly(dimethylsiloxane) (PDMS) based microfluidic devices. *Microfluid Nanofluid* 2009;7:291-306.

[37] Doherty EAS, Meagher RJ, Albarghouthi MN, Barron AE. Microchannel wall coatings for protein separations by capillary and chip electrophoresis. *Electrophoresis* 2003;24:34-54.

[38] Makamba H, Kim JH, Lim K, Park N, Hahn JH. Surface modification of poly(dimethylsiloxane) microchannels. *Electrophoresis* 2003;24:3607-19.

[39] Kirby BJ, Wheeler AR, Zare RN, Fruetel JA, Shepodd TJ. Programmable modification of cell adhesion and zeta potential in silica microchips. *Lab Chip* 2003;3:5-10.

[40] Muck A, Svatos A. Chemical modification of polymeric microchip devices. *Talanta* 2007;74:333-41.

[41] Chen Y, Zhang LY, Chen G. Fabrication, modification, and application of department of analytical poly(methyl methacrylate) microfluidic chips. *Electrophoresis* 2008;29:1801-14.

[42] Liu JK, Pan T, Woolley AT, Lee ML. Surface-modified poly(methyl methacrylate) capillary electrophoresis microchips for protein and peptide analysis. *Anal Chem* 2004;76:6948-55.

[43] Harrison DJ, Manz A, Fan ZH, Ludi H, Widmer HM. Capillary Electrophoresis and Sample Injection Systems Integrated on a Planar Glass Chip. *Anal Chem* 1992;64:1926-32.

[44] Xia YN, Whitesides GM. Soft lithography. *Annu Rev Mater Sci* 1998;28:153-84.

[45] McDonald JC, Duffy DC, Anderson JR, Chiu DT, Wu HK, Schueller OJA, et al. Fabrication of microfluidic systems in poly(dimethylsiloxane). *Electrophoresis* 2000;21:27-40.

[46] Metz S, Holzer R, Renaud P. Polyimide-based microfluidic devices. *Lab Chip* 2001;1:29-34.

[47] Rossier J, Reymond F, Michel PE. Polymer microfluidic chips for electrochemical and biochemical analyses. *Electrophoresis* 2002;23:858-67.

[48] Fiorini GS, Chiu DT. Disposable microfluidic devices: fabrication, function, and application. *Biotechniques* 2005;38:429-46.

[49] Stachowiak TB, Svec F, Frechet JMJ. Patternable protein resistant surfaces for multifunctional microfluidic devices via surface hydrophilization of porous polymer monoliths using photografting. *Chem Mater* 2006;18:5950-7.

[50] Hu SW, Ren XQ, Bachman M, Sims CE, Li GP, Allbritton N. Surface modification of poly(dimethylsiloxane) microfluidic devices by ultraviolet polymer grafting. *Anal Chem* 2002;74:4117-23.

[51] Sakai-Kato K, Kato M, Ishihara K, Toyo'oka T. An enzyme-immobilization method for integration of biofunctions on a microchip using a water-soluble amphiphilic phospholipid polymer having a reacting group. *Lab Chip* 2004;4:4-6.

[52] Nishizawa K, Konno T, Takai M, Ishihara K. Bioconjugated phospholipid polymer biointerface for enzyme-linked immunosorbent assay. *Biomacromolecules* 2008;9:403-7.

[53] Sibarani J, Takai M, Ishihara K. Surface modification on microfluidic devices with 2-methacryloyloxyethyl phosphorylcholine polymers for reducing unfavorable protein adsorption. *Colloid Surf B-Biointerfaces* 2007;54:88-93.

[54] Oki A, Takai M, Ogawa H, Takamura Y, Fukasawa T, Kikuchi J, et al. Healthcare chip for checking health condition from analysis of trace blood collected by painless needle. *Jpn J Appl Phys* 1 2003;42:3722-7.

[55] Oki A, Adachi S, Takamura Y, Ishihara K, Ogawa H, Ito Y, et al. Electroosmosis injection of blood serum into biocompatible microcapillary chip fabricated on quartz plate. *Electrophoresis* 2001;22:341-7.

[56] Bi H, Zhong W, Meng S, Kong J, Yang P, Liu B. Construction of a biomimetic surface on microfluidic chips for biofouling resistance. *Anal Chem* 2006;78:3399-405.

[57] Nishizawa K, Takai M, Ishihara K. Electrosprayed phospholipid polymer surface for enhanced sensitive ELISA chip. In: Hasebe S, editor. *10th International Conference on*

Miniaturized Systems for Chemistry and Life Sciences; 2006: Tokyo: Jpn. Assoc. Inc.; 2006. p. 335-7.

[58] Nishizawa K, Konno T, Takai M, Ishihara K. Nanostructured phospholipid biointerface for immunoassay microchip integrated with plasma separation membrane. In: Viovy JL, Tabeling P, Descroix S, Malaquin L, editors. 11th International Conference on Miniaturized Systems for Chemistry and Life Sciences; 2007: San Diego: *Chem. Biol. Microsys. Soc.;* 2007. p. 1291-3.

[59] Xu Y, Takai M, Konno T, Ishihara K. Microfluidic flow control on charged phospholipid polymer interface. *Lab Chip* 2007;7:199-206.

[60] Xu Y, Takai M, Ishihara K. Suppression of Protein Adsorption on a Charged Phospholipid Polymer Interface. *Biomacromolecules* 2009;10:267-74.

[61] Xu Y, Takai M, Ishihara K. Protein adsorption and cell adhesion on cationic, neutral, and anionic 2-methacryloyloxyethyl phosphorylcholine copolymer surfaces. *Biomaterials* 2009;30:4930-8.

[62] Jang K, Sato K, Mawatari K, Konno T, Ishihara K, Kitamori T. Surface modification by 2-methacryloyloxyethyl phosphorylcholine coupled to a photolabile linker for cell micropatterning. *Biomaterials* 2009;30:1413-20.

[63] Goto M, Tsukahara T, Sato K, Konno T, Ishihara K, Sato K, et al. Nanometer-scale patterned surfaces for control of cell adhesion. *Anal Sci* 2007;23:245-7.

[64] Hamada Y, Ono T, Akagi T, Ishihara K, Ichiki T. Photochemical oxidation of poly(dimethylsiloxane) surface and subsequent coating with biomimetic phosphorylcholine polymer. *J Photopolym Sci Tec* 2007;20:245-9.

[65] Hoshino T, Konno T, Ishihara K, Morishima K. Live-Cell-Driven Insertion of a Nanoneedle. *Jpn J Appl Phys* 2009;48:-.

[66] Goda T, Konno T, Takai M, Moro T, Ishihara K. Biomimetic phosphorylcholine polymer grafting from polydimethylsiloxane surface using photo-induced polymerization. *Biomaterials* 2006;27:5151-60.

[67] Goda T, Matsuno R, Konno T, Takai M, Ishihara K. Photografting of 2-methacryloyloxyethyl phosphorylcholine from polydimethylsiloxane: Tunable protein repellency and lubrication property. *Colloid Surf B-Biointerfaces* 2008;63:64-72.

[68] Singer SJ, Nicolson GL. Fluid Mosaic Model of Structure of Cell-Membranes. *Science* 1972;175:720-&.

[69] Nakai S, Nakaya T, Imoto M. Polymeric Phospholipid Analog, .10. Synthesis and Polymerization of 2-(Methacryloyloxy)Ethyl 2-Aminoethyl Hydrogen Phosphate. *Makromol Chem* 1977;178:2963-7.

[70] Kadoma Y, Nakabayashi N, Masuhara E, Yamauchi J. Synthesis and Hemolysis Test of Polymer Containing Phosphorylcholine Groups. *Kobunshi Ronbunshu* 1978;35:423-7.

[71] Hupfer B, Ringsdorf H, Schupp H. Polyreactions in Oriented Systems .21. Polymeric Phospholipid Monolayers. *Macromol Chem Phys* 1981;182:247-53.

[72] Durrani AA, Hayward JA, Chapman D. Biomembranes as Models for Polymer Surfaces .2. The Syntheses of Reactive Species for Covalent Coupling of Phosphorylcholine to Polymer Surfaces. *Biomaterials* 1986;7:121-5.

[73] Ringsdorf H, Schlarb B. Preparation and Characterization of Unsymmetrical Liposomes in a Net. *Makromol Chem* 1988;189:299-315.

[74] Hall B, Bird RL, Kojima M, Chapman D. Biomembranes as Models for Polymer Surfaces .5. Thromboelastographic Studies of Polymeric Lipids and Polyesters. *Biomaterials* 1989;10:219-24.

[75] Chapman D. Biomembranes and New Hemocompatible Materials. *Langmuir* 1993;9:39-45.

[76] Marra KG, Winger TM, Hanson SR, Chaikof EL. Cytomimetic biomaterials .1. In-situ polymerization of phospholipids on an alkylated surface. *Macromolecules* 1997;30:6483-8.

[77] Lewis AL. Phosphorylcholine-based polymers and their use in the prevention of biofouling. *Colloid Surf B-Biointerfaces* 2000;18:261-75.

[78] Matsuda T, Nagase J, Ghoda A, Hirano Y, Kidoaki S, Nakayama Y. Phosphorylcholine-endcapped oligomer and block co-oligomer and surface biological reactivity. *Biomaterials* 2003;24:4517-27.

[79] Ishihara K, Ueda T, Nakabayashi N. Preparation of Phospholipid Polymers and Their Properties as Polymer Hydrogel Membranes. *Polym J* 1990;22:355-60.

[80] Ueda T, Oshida H, Kurita K, Ishihara K, Nakabayashi N. Preparation of 2-Methacryloyloxyethyl Phosphorylcholine Copolymers with Alkyl Methacrylates and Their Blood Compatibility. *Polym J* 1992;24:1259-69.

[81] Kojima M, Ishihara K, Watanabe A, Nakabayashi N. Interaction between Phospholipids and Biocompatible Polymers Containing a Phosphorylcholine Moiety. *Biomaterials* 1991;12:121-4.

[82] Ma YH, Tang YQ, Billingham NC, Armes SP, Lewis AL. Synthesis of biocompatible, stimuli-responsive, physical gels based on ABA triblock copolymers. *Biomacromolecules* 2003;4:864-8.

[83] Seo JH, Matsuno R, Takai M, Ishihara K. Cell adhesion on phase-separated surface of block copolymer composed of poly(2-methacryloyloxyethyl phosphorylcholine) and poly(dimethylsiloxane). *Biomaterials* 2009;30:5330-40.

[84] Ishihara K, Tsuji T, Kurosaki T, Nakabayashi N. Hemocompatibility on Graft-Copolymers Composed of Poly(2-Methacryloyloxyethyl Phosphorylcholine) Side-Chain and Poly(N-Butyl Methacrylate) Backbone. *J Biomed Mater Res* 1994;28:225-32.

[85] Lewis AL, Berwick J, Davies MC, Roberts CJ, Wang JH, Small S, et al. Synthesis and characterisation of cationically modified phospholipid polymers. *Biomaterials* 2004;25:3099-108.

[86] Konno T, Ishihara K. Temporal and spatially controllable cell encapsulation using a water-soluble phospholipid polymer with phenylboronic acid moiety. *Biomaterials* 2007;28:1770-7.

[87] Xu Y, Sato K, Mawatari K, Konno T, Jang K, Ishihara K, et al. A Microfluidic Hydrogel Capable of Cell Preservation without Perfusion Culture under Cell-Based Assay Conditions. *Adv Mater* 2010;22:3017-21.

[88] Xu Y, Jang K, Konno T, Ishihara K, Mawatari K, Kitamori T. The biological performance of cell-containing phospholipid polymer hydrogels in bulk and microscale form. *Biomaterials* 2010;31:8839-46.

[89] Iwasaki Y, Yamasaki A, Ishihara K. Platelet compatible blood filtration fabrics using a phosphorylcholine polymer having high surface mobility. *Biomaterials* 2003;24:3599-604.

[90] Kyomoto M, Moro T, Saiga K, Miyaji F, Kawaguchi H, Takatori Y, et al. Lubricity and stability of poly(2-methacryloyloxyethyl phosphorylcholine) polymer layer on Co-Cr-Mo surface for hemi-arthroplasty to prevent degeneration of articular cartilage. *Biomaterials* 2010;31:658-68.

[91] Hasegawa T, Iwasaki Y, Ishihara K. Preparation and performance of protein-adsorption-resistant asymmetric porous membrane composed of polysulfone/ phospholipid polymer blend. *Biomaterials* 2001;22:243-51.

[92] Sawada SI, Iwasaki Y, Nakabayashi N, Ishihara K. Stress response of adherent cells on a polymer blend surface composed of a segmented polyurethane and MPC copolymers. *J Biomed Mater Res A* 2006;79A:476-84.

[93] Iwasaki Y, Nakabayashi N, Ishihara K. Preservation of platelet function on 2-methacryloyloxyethyl phosphorylcholine-graft polymer as compared to various water-soluble graft polymers. *J Biomed Mater* Res 2001;57:72-8.

[94] Kyomoto M, Moro T, Iwasaki Y, Miyaji F, Kawaguchi H, Takatori Y, et al. Superlubricious surface mimicking articular cartilage by grafting poly(2-methacryloyloxyethyl phosphorylcholine) on orthopaedic metal bearings. *J Biomed Mater Res A* 2009;91A:730-41.

[95] Park J, Kurosawa S, Takai M, Ishihara K. Antibody immobilization to phospholipid polymer layer on gold substrate of quartz crystal microbalance immunosensor. *Colloid Surf B-Biointerfaces* 2007;55:164-72.

[96] Shimada T, Ueda M, Jinno H, Chiba N, Wada M, Watanabe J, et al. Development of Targeted Therapy with Paclitaxel Incorporated into EGF-conjugated Nanoparticles. *Anticancer Res* 2009;29:1009-14.

[97] Ye SH, Watanabe J, Takai M, Iwasaki Y, Ishihara K. High functional hollow fiber membrane modified with phospholipid polymers for a liver assist bioreactor. *Biomaterials* 2006;27:1955-62.

[98] Fukazawa K, Ishihara K. Fabrication of a cell-adhesive protein imprinting surface with an artificial cell membrane structure for cell capturing. *Biosens Bioelectron* 2009;25:609-14.

[99] Moro T, Takatori Y, Ishihara K, Konno T, Takigawa Y, Matsushita T, et al. Surface grafting of artificial joints with a biocompatible polymer for preventing periprosthetic osteolysis. *Nat Mater* 2004;3:829-36.

[100] Ye SH, Johnson CA, Woolley JR, Snyder TA, Gamble LJ, Wagner WR. Covalent surface modification of a titanium alloy with a phosphorylcholine-containing copolymer for reduced thrombogenicity in cardiovascular devices. *J Biomed Mater Res A* 2009;91A:18-28.

[101] Nam K, Watanabe J, Ishihara K. Modeling of swelling and drug release behavior of spontaneously forming hydrogels composed of phospholipid polymers. *Int J Pharm* 2004;275:259-69.

[102] Ahmed M, Bhuchar N, Ishihara K, Narain R. Well-Controlled Cationic Water-Soluble Phospholipid Polymer-DNA Nanocomplexes for Gene Delivery. *Bioconjugate Chem* 2011;22:1228-38.

[103] Choi J, Konno T, Takai M, Ishihara K. Regulation of cell proliferation by multi-layered phospholipid polymer hydrogel coatings through controlled release of paclitaxel. *Biomaterials* 2012;33:954-61.

[104] Kameyama N, Matsuda S, Itano O, Ito A, Konno T, Arai T, et al. Photodynamic Therapy Using an Anti-EGF Receptor Antibody Complexed with Verteporfin Nanoparticles: A Proof of Concept Study. *Cancer Biother Radio* 2011;26:697-704.

[105] Watanabe J, Eriguchi T, Ishihara K. Cell adhesion and morphology in porous scaffold based on enantiomeric poly(lactic acid) graft-type phospholipid polymers. *Biomacromolecules* 2002;3:1375-83.

[106] Il Kim H, Takai M, Ishihara K. Bioabsorbable Material-Containing Phosphorylcholine Group-Rich Surfaces for Temporary Scaffolding of the Vessel Wall. *Tissue Eng Pt C-Meth* 2009;15:125-33.

[107] Park JW, Kurosawa S, Aizawa H, Goda Y, Takai M, Ishihara K. Piezoelectric immunosensor for bisphenol A based on signal enhancing step with 2-methacrolyloxyethyl phosphorylcholine polymeric nanoparticle. *Analyst* 2006;131:155-62.

[108] Chu MX, Kudo H, Shirai T, Miyajima K, Saito H, Morimoto N, et al. A soft and flexible biosensor using a phospholipid polymer for continuous glucose monitoring. *Biomed Microdevices* 2009;11:837-42.

[109] Jacobson SC, Moore AW, Ramsey JM. Fused Quartz Substrates for Microchip Electrophoresis. *Anal Chem* 1995;67:2059-63.

[110] Koutny LB, Schmalzing D, Taylor TA, Fuchs M. Microchip electrophoretic immunoassay for serum cortisol. *Anal Chem* 1996;68:18-22.

[111] Woolley AT, Lao KQ, Glazer AN, Mathies RA. Capillary electrophoresis chips with integrated electrochemical detection. *Anal Chem* 1998;70:684-8.

[112] He B, Tan L, Regnier F. Microfabricated filters for microfluidic analytical systems. *Anal Chem* 1999;71:1464-8.

[113] Tian HJ, Landers JP. Hydroxyethylcellulose as an effective polymer network for DNA analysis in uncoated glass microchips: optimization and application to mutation detection via heteroduplex analysis. *Anal Biochem* 2002;309:212-23.

[114] Gau H, Herminghaus S, Lenz P, Lipowsky R. Liquid morphologies on structured surfaces: From microchannels to microchips. *Science* 1999;283:46-9.

[115] Gallardo BS, Gupta VK, Eagerton FD, Jong LI, Craig VS, Shah RR, et al. ELectrochemical principles for active control of liquids on submillimeter scales. *Science* 1999;283:57-60.

[116] Kingshott P, Griesser HJ. Surfaces that resist bioadhesion. *Curr Opin Solid St M* 1999;4:403-12.

[117] Ostuni E, Chapman RG, Liang MN, Meluleni G, Pier G, Ingber DE, et al. Self-assembled monolayers that resist the adsorption of proteins and the adhesion of bacterial and mammalian cells. *Langmuir* 2001;17:6336-43.

[118] Sibarani J, Konno T, Takai M, Ishihara K. Surface modification by grafting with biocompatible 2-methacryloyloxyethyl phosphorylcholine for microfluidic devices. *Key Eng Mater* 2007;342-343:789-92.

[119] Nishizawa K, Konno T, Takai M, Ishihara K. Microchip immunoassay using high density bioconjugation on the phospholipid polymer interface. *Trans Mater Res Soc Jpn* 2008;33:787-90.

[120] Nishizawa K, Konno T, Takai M, Ishihara K. Stabilization of bioconjugated phospholipid polymer nanometer scaled structures for highly-sensitive immunoassay. *Trans Mater Res Soc Jpn* 2009;34:201-4.

[121] Erim FB, Cifuentes A, Poppe H, Kraak JC. Performance of a Physically Adsorbed High-Molecular-Mass Polyethyleneimine Layer as Coating for the Separation of Basic-Proteins and Peptides by Capillary Electrophoresis. *J Chromatogr A* 1995;708:356-61.

[122] Katayama H, Ishihama Y, Asakawa N. Stable cationic capillary coating with successive multiple ionic polymer layers for capillary electrophoresis. *Anal Chem* 1998;70:5272-7.

[123] Pasche S, Voros J, Griesser HJ, Spencer ND, Textor M. Effects of ionic strength and surface charge on protein adsorption at PEGylated surfaces. *J Phys Chem B* 2005;109:17545-52.

[124] Lund M, Akesson T, Jonsson B. Enhanced protein adsorption due to charge regulation. *Langmuir* 2005;21:8385-8.

[125] Holmlin RE, Chen XX, Chapman RG, Takayama S, Whitesides GM. Zwitterionic SAMs that resist nonspecific adsorption of protein from aqueous buffer. *Langmuir* 2001;17:2841-50.

[126] Lewis AL, Cumming ZL, Goreish HH, Kirkwood LC, Tolhurst LA, Stratford PW. Crosslinkable coatings from phosphorylcholine-based polymers. *Biomaterials* 2001;22:99-111.

[127] Clarke S, Davies MC, Roberts CJ, Tendler SJB, Williams PM, O'Byrne V, et al. Surface mobility of 2-methacryloyloxyethyl phosphorylcholine-co-lauryl methacrylate polymers. *Langmuir* 2000;16:5116-22.

[128] Feng W, Zhu SP, Ishihara K, Brash JL. Adsorption of fibrinogen and lysozyme on silicon grafted with poly(2-methacryloyloxyethyl phosphorylcholine) via surface-initiated atom transfer radical polymerization. *Langmuir* 2005;21:5980-7.

[129] Futamura K, Matsuno R, Konno T, Takai M, Ishihara K. Rapid development of hydrophilicity and protein adsorption resistance by polymer surfaces bearing phosphorylcholine and naphthalene groups. *Langmuir* 2008;24:10340-4.

[130] Kishida A, Iwata H, Tamada Y, Ikada Y. Cell Behavior on Polymer Surfaces Grafted with Nonionic and Ionic Monomers. *Biomaterials* 1991;12:786-92.

[131] Ishihara K, Nomura H, Mihara T, Kurita K, Iwasaki Y, Nakabayashi N. Why do phospholipid polymers reduce protein adsorption? *J Biomed Mater Res* 1998;39:323-30.

[132] Ishihara K, Oshida H, Endo Y, Ueda T, Watanabe A, Nakabayashi N. Hemocompatibility of Human Whole-Blood on Polymers with a Phospholipid Polar Group and Its Mechanism. *J Biomed Mater Res* 1992;26:1543-52.

[133] Morisaku T, Watanabe J, Konno T, Takai M, Ishihara K. Hydration of phosphorylcholine groups attached to highly swollen polymer hydrogels studied by thermal analysis. *Polymer* 2008;49:4652-7.

[134] Sheng Q, Schulten K, Pidgeon C. Molecular-Dynamics Simulation of Immobilized Artificial Membranes. *J Phys Chem-Us* 1995;99:11018-27.

[135] Kitano H, Imai M, Mori T, Gemmei-Ide M, Yokoyama Y, Ishihara K. Structure of water in the vicinity of phospholipid analogue copolymers as studied by vibrational spectroscopy. *Langmuir* 2003;19:10260-6.

[136] Palmer RR, Lewis AL, Kirkwood LC, Rose SF, Lloyd AW, Vick TA, et al. Biological evaluation and drug delivery application of cationically modified phospholipid polymers. *Biomaterials* 2004;25:4785-96.

[137] Stroock AD, Weck M, Chiu DT, Huck WTS, Kenis PJA, Ismagilov RF, et al. Patterning electro-osmotic flow with patterned surface charge. *Phys Rev Lett* 2000;84:3314-7.

[138] Szekely L, Guttman A. New advances in microchip fabrication for electrochromatography. *Electrophoresis* 2005;26:4590-604.

[139] Futterer C, Minc N, Bormuth V, Codarbox JH, Laval P, Rossier J, et al. Injection and flow control system for microchannels. *Lab Chip 2004;4:351-6.*

[140] Kohlheyer D, Besselink GAJ, Lammertink RGH, Schlautmann S, Unnikrishnan S, Schasfoort RBM. Electro-osmotically controllable multi-flow microreactor. *Microfluid Nanofluid* 2005;1:242-8.

[141] Moorthy J, Khoury C, Moore JS, Beebe DJ. Active control of electroosmotic flow in microchannels using light. *Sensor Actuat B-Chem* 2001;75:223-9.

[142] Schasfoort RBM, Schlautmann S, Hendrikse L, van den Berg A. Field-effect flow control for microfabricated fluidic networks. *Science* 1999;286:942-5.

[143] Dittmann MM, Rozing GP. Capillary electrochromatography: Investigation of the influence of mobile phase and stationary phase properties on electroosmotic velocity, retention, and selectivity. *J Microcolumn Sep* 1997;9:399-408.

[144] Kirby BJ, Hasselbrink EF. Zeta potential of microfluidic substrates: 1. Theory, experimental techniques, and effects on separations. *Electrophoresis* 2004;25:187-202.

[145] Konig S, Welsch T. Moderation of the electroosmotic flow in capillary electrophoresis by chemical modification of the capillary surface with tentacle-like oligourethanes. *J Chromatogr A* 2000;894:79-88.

[146] Shimizu K, Nagase Y, Takai M, Ishihara K. Biofouling Control by Phospholipid Polymer on Microchannel DNA Separation. *Trans Mater Res Soc Jpn* 2006;31:1069-72.

[147] Takai M, Shimizu K, Nagase Y, Ishihara K. DNA separation on nanochannel with phospholipid polymers. In: Hasebe S, editor. 10th International Conference on Miniaturized Systems for Chemistry and Life Sciences; 2006: Tokyo: *Jpn. Assoc. Inc.;* 2006. p. 885-7.

[148] Xu Y, Takai M, Ishihara K. Phospholipid Polymer Biointerfaces for Lab-on-a-Chip Devices. *Ann Biomed Eng* 2010;38:1938-53.

[149] Futamura K, Matsumoto A, Sakata T, Ishihara K, Miyahara Y. Phospholipid polymer-modified field effect transistor for sensitive detection of biomolecular recognition. In: Kim TS, Lee Y, Chung T, Jeon NL, Lee S, Suh K, et al., editors. 13th International Conference on Miniaturized Systems for Chemistry and Life Sciences; 2009: San Diego: *Chem. Biol. Microsys. Soc.;* 2009. p. 1557-9.

In: Silane: Chemistry, Applications and Performance
Editors: Katsuo Moriguchi and Susumu Utagawa
ISBN: 978-1-62257-432-2

Chapter 2

Review of Status, Trends, and Challenges in Working with Silane and Functional Silanes

Pablo Mancheno-Posso[1], Roy F. Dittler[1], Deanna Lewis[2], Polly Juang[3], Xiaoyang Ji[4], Xinhai Xu[5] and David C. Lynch[6]

[1]Depts. of Chemical and Environmental Engr., US
[2]Mel and Enid Zuckerman College of Public Health, US
[3]Agriculture and Biosystems Engr., US
[4]Materials Science and Engr.
[5]Aerospace and Mechanical Engr., US
[6]Mining Engr. and Materials Science and Engr.
University of Arizona, Tucson, Arizona, US

Abstract

In this chapter the status, trends, and challenges of working with silane and functional silanes are examined. The subject is far too vast to be covered in a single chapter, and much of what has been accomplished is proprietary and not available for review.

Thus the focus of this review is on standard practices and advancements published in the literature over the last twenty years; beginning with review of silane applications.Processes for formation and purification of silanes are reviewed with emphasis on the Direct, Siemens, Union Carbide/REC, Schmidt, and Ethyl/MEMC processes.

The review continues with examination of the role silanes play in the processing of electronic chips, and analysis of hazards and public health issues one must be aware of when working with these compounds. The review concludes with examination of thermo-chemical data for silanes.

INTRODUCTION

The term silane, once limited to silicon hydrides, has grown to include functional silanes, where one or more hydrogen atoms are replaced by a radical such as Cl, F, CH_3, etc. That practice has led some to extend naming of silanes to the point where silicon tetrachloride is referred to by some as tetrachlorosilane, a practice the authors reject, confining the definition of silanes to silicon hydrides with or without radicals. Silanes find extensive use as precursors in the chemical industry, as reagent in the production of high purity silicon, as surface modifiers for adhesion of various coating, as the transition compound from organic to inorganic materials, as precursors in producing a range of products, including silicones, and many other uses presented in Table 1 [1].

Table 1. Classification and Uses for Silanes (largely based on ref. [1])

Inorganic Silanes • Precursor to production of organosilanes • Precursor to production of electronic and solar grade silicons • Reducing agent • Used to produce integrated circuit (IC) devices • Used to produce electrical isolation for IC interconnects
Aminosilanes • Promotes adhesion, serves as coupling agent, and is used as resin additives • Serves as coupling agent between fillers and fibers in composites by promoting chemical bonding between phases, also used as surface pretreatment for fillers and reinforcement fibers • Finds uses with epoxies, phenolics, melamines, nylons, PVC, acrylics, polyolefins, polyurethanes, and nitrile rubbers
Vinyl Silanes • Promotes crosslinking and serves as oupling agent • Used with sulfur and peroxide cured rubber, polyester, polyolefins, styrenics, and acrylics • Used with ethylene to copolmerize for moisture induced coupling of polyethylene • Used with polyethylene for moisture induced coupling • Surface modifier for mineral fillers in plastics
Epoxy Silanes • Adhesion promoter • Surface treatment for fillers and reinforcement fibers
Methacryl Silanes • Promotes adhesion and serves as coupling agent including moisture cross linking of acrylics • Used for free-radical, crosslinked polyester, rubber, poly(olefins), styrenics, and acrylics • Coupling agent between fillers or fiberglass to resins
Alkylsilanes • Hydrophobic surface treatment • Silicone synthesis
Phenyl Silanes • Hydrophobic treatment of both surfaces and silane coupling agents • Improves thermal stability of other silanes as additive • Use in synthesis of silanes and siloxanes
Organochlorosilanes • Use in synthesis of silanes and siloxanes • Protecting agents for intermediates in pharmaceutical syntheses (methylchlorosilanes)

Figure 1. Fireball from detonation of gas cylinder of SiH_4 (reproduced with permission from the University of Arizona).

Silanes impact our lives in many ways given their role as precursors. One example is the halogenation of any of the basic silane series, Si_nH_{2n+2} where n has a value of 1 to at least 6. The resulting halogenated silane is reacted with ammonia to form silylamines and silazanes that are then hydrolyzed by water to form siloxanes that consist of R_2SiO groups, where R is a hydrogen atom or a hydrocarbon group. Siloxanes have found use in cosmetics, deodorants, soap, anti-foaming agents, water repellants, and food additives [2].

At this point it is advisable to inject a few words regarding nomenclature for silanes. The word *silane* is used here to define linear compounds; silane SiH_4, disilane Si_2H_6, trisilane Si_3H_8,...etc. Silanes with a SiH_3 branch, or radical, contain the prefix silyl in the naming of the compound.An example is 2-silytrisilane which has the formula $Si_3H_7(SiH_3)$ with the silyl radical bonded to the second silicon atom in the chain. Silanes form circular structures which as a class are referred to as cyclosilanes. Silanes with a circular hydrocarbon radical are part of the aryl group of silanes. Whereas silanes with linear hydrocarbon radicals (C_nH_{2n+1}) are part of the alkyl group of silanes. When it comes to writing the formulas for functional silanes there appears to be few rules. An example is the chemical formula for trichlorosilane, also known as TCS, which is written in the literature as HCl_3Si, $SiHCl_3$, or as $HSiCl_3$.

The review that follows is focused on methods for formation and purification of silanes and their uses, including their role in the electronics industry. Following that material is an analysis of health and safety issues. The danger in dealing with silane (SiH_4) cannot be understated. Silane explosions have led to fatalities. An example of the danger is shown in Figure 1 where a gas cylinder of SiH_4 with regulator left unattended for an extensive period was destroyed by inducing detonation. That means of disposal was considered the safest approach, as energy to produce a detonation is so low that some explosions are attributed to unknown ignition sources.

SILANE APPLICATIONS

Silanes have a wide variety of uses from the production of penicillin drugs to combustion initiation in supersonic ramjets [3 and 4]. Additionally, silanes have many different types of applications: coupling agents, adhesion promoters, cross-linkers, hydrophobic and oleophobic agents, dispersing agents, silicate stabilizers, and water scavengers [5]. These applications usually use organofunctional alkoxysilanes which are defined as silanes containing at least one carbon-silicon bond. i.e. example (CH_3-Si). Organofunctional silanes are stable, non-polar and hydrophobic structures [6]. When using silanes with polymers, the distinction of organofunctional and non-organofunctional silanes is important to create the desired effect of a silane product. Organofunctional silanes have a reactive organic group with structural, thermal, chemical, and solubility properties that are designed to match the properties of the polymer being used with the silane.

Table 2. Characteristics of Organosilanes [7]

Organosilanes R-$SiH_x(OMe)_{3-x}$	
R Group	Characteristics of R Group
Me	Hydrophobic, Organophilic
Ph	Hydrophobic, Organophilic, Thermal Stability
i-Bu	Hydrophobic, Organophilic
Octyl	Hydrophobic, Organophilic
-$NH(CH_2)_3NH_2$	Hydrophilic, Organoreactive
Epoxy	Hydrophilic, Organoreactive
Methacryl	Hydrophobic, Organoreactive

Table 2 shows some organofunctional silanes and their characteristics. Non-organofunctional silanes are used to modify the characteristics of the polymer used with the silane to include properties such as hydrophobicity, decreased surface energy and compatibility with other organic materials. A few examples of non-organofunctional silanes are tetraethoxysilane, methyltrimeth-oxysilane, dimethyldimethoxysilane, propyltrimethoxy silane, isobutyltrimethosxysilane, phenyltrimethoxysilane and n-octyltriethoxysilane [7].

Functional silanes find use as:

- Coupling Agents

Coupling agents bind organic polymers to fillers such as fiberglass/silica or mineral fillers (talc, mica, clay, and wollastonite) [5]. As a coupling agent, silanes help to improve bonding and mixing as well as to increase the matrix strength between these polymers and fillers [5].

Fillers enhance the functionality of the polymers, usually rubber or plastic, to form composite materials to strengthen it by means of stress transfer. The smaller the filler size, the stronger the composite material as the surface area of the filler increases [8].

Silanes as coupling agents find use in automobile body parts, soles of shoes, printed circuit boards, plastic pipes, wind turbine blades, storage tanks, tires, shower stalls, satellite dishes, construction tools, and coating of wires and cables [5].

- Adhesion Promoters

Silanes facilitate adhesion on substrates (usually glass, metal, and stone), and are often used as primers in paints and glues. Silanes used as coupling agents are also efficient adhesion promoters [5].

Silanes are used as adhesion promoters because most of them have properties that are moisture, temperature and chemical resistant. Many types of coatings and surfaces have adhesion problems when exposed to moisture or salt spray in which the addition of an adhesion promoter will resolve the problem [9]. Silanes as adhesion promoters find use in paint primers, inks, glue primers, sealants, adhesives, and coatings [10].

- Cross-Linkers

Some silanes can cross-link an organic polymer to siliceous or mineral fillers for heat, water, tear, abrasion and stretch resistance. Cross-linking silanes help to improve mixing, bonding and matrix strength between polymers (ie. polyethylene, acrylates, polyethers, polyurethanes and polyesters) and organic fillers (ie. acrylics and urethanes) [9 and 5]. Cross-linking silanes find use in paints, coatings and adhesives [11].

- Hydrophobic and Oleophobic Agents

Organofunctional silanes with organic groups have hydrophobic and oleophobic characteristics when applied to inorganic surfaces. These hydrophobic silanes are used mostly in construction applications such as bridge and building decks. As dispersing agents, these silanes can hydrophobe inorganic powders to create a free flowing mix that can be dispersed or integrated into organic liquids and polymers [5]. Hydrophobic silanes find use in working with concrete, clay, brick, quartz, talc, limestone, kaolin, mica, gypsum, marble,and perlite. Oleophobic silanes find uses with the same materials[12 and 13].

- Water Scavengers

Organofuntional silanes quickly react with water molecules, hydrolyzing alkoxy groups attached to the silane, to convert to alcohol molecules. These silanes are often used as water scavengers in moisture sensitive products. The effect of the water scavenging helps to stabilize a material and to prevent premature curing as well as uniform curing [5]. Organofuntional silanes find use in paints, sealants, adhesives, pipes and cables, metal treatments, and coatings [9].

- Other Uses

Silanes may also be used as silicate stabilizers especially in engine coolant to inhibit silicate conglomeration [5]. Some other examples of common items that include the usage of silanes include pharmaceutical manufacturing (penicillin), silicon wafers (photovoltaic panels), electronics (chips), combustion initiation (supersonic combustion ramjets) and dental bonding (porcelain fillings).

Methods for Producing Silanes

There are numerous references in the literature [14 and 15] as to methods for producing silanes, and more held proprietary by producers. The discussion that follows on producing silanes is focused on the Direct Process, and processes that utilize silanes in the purification of silicon for the electronics and photovoltaic industries because of their commercial importance. That material is followed by a short discussion of other methods reported in the literature to produce silanes. That material is presented in tabular form, and includes references for those who wish to explore a process in greater depth.

The Direct Process (also referred to as the Rochow Process) is critical in producing organosilanes. Earlier the decision was made to limit the discussion in this chapter to silanes with at least one hydrogen atom bonded to silicon to reduce the scope of this chapter. An exception is made here for the Direct Process because of its commercial importance. That process produces methychlorosilanes, but the primary product is Me_2SiCl_2. By the limiting definition adopted, the primary product is not a silane. The Me_2SiCl_2 and other methylchlorosilanes produced through the Direct Process are used to produce many of the commercially available organosilanes, and thus the process is reviewed here.

Direct Process

Eugene Rochow first reported the synthesis of hydrocarbon-silicon halides in 1941 [16]. Rochow demonstrated that a solid mixture containing silicon could be reacted with methyl chloride vapor at temperatures between 473 and 773 K using a copper catalyst. The new method created an alternative to the already existent techniques, and significantly improved the economics in producing organosilanes.

The Direct Process on a commercial scale takes place in a fluidized-bed reactor. The bed is composed of silicon particulate and is fluidized by a gaseous alkyl halide, or its mixture with an inert gas. Methyl chloride is the preferred alkyl halide [17]. Commercial reactors can be 15 m tall, with a diameter of 1.5 to 5 m, and can be loaded with 100 tons of solid mixture [18]. The Direct Process involves the general reaction [14]:

$$2RCl(g) + Si \rightarrow R_2SiCl_2 \tag{1}$$

where R represents an organic functional group. Silicon reacts with organic chlorides to produce $(C_nH_{2n+1})_2SiCl_2$. Those compounds with one or more alkyl groups can be generated, but the methyl radical (Me) is the norm for the Direct Process. On an industrial scale copper, copper oxide, and copper chloride are used as catalysts, and the reactor temperature varies between 543 and 593 K [17]. The actual process consists of several reactions in parallel, in which Me_2SiCl_2, $MeSiHCl_2$, Me_3SiCl, and $MeSiCl_3$ are produced along with several low boilers (Me_4Si, $SiHCl_3$, and Me_2SiHCl). The process also generates disilane and other polysilanes as high boilers, vents CH_4, ethane and other hydrocarbon gases, and leaves a solid residue of Si, C, $CuCl_2$, CuCl, copper silicides, and other metal chlorides and silicides. Ten percent of the silicon charged to the process is lost in the solid residue. With an appropriate catalyst, the selectivity towards Me_2SiCl_2 is commonly between 80 and 90% [19]. The selectivity towards a specific product depends on the catalyst, nature of the reactants, temperature, pressure, and technological approach [20 and 21]. Selectivity is critical as $MeSiCl_3$ is difficult to remove from the product stream, and hinders formation of linear polymers in forming silicones.

While the chemical reactions for the Direct Process are known, the reaction mechanism is not well established. While the reactivity and selectivity of the process is reproducible with purified Si, the presence of different silicides in silicon has been shown to significantly impact both reactivity and selectivity [22 - 24]. While silicon with Fe and Fe-Al silicides tend to have the highest and most stable reactivity, they have lower selectivity. Silicon containing Ca-Fe-Al silicides tend to have the highest selectivity, but lower reactivity. Depending on the presence of different silicides in the silicon the reactivity and selectivity change with the extent a particle is reacted. Given the variable nature of silicides in metallurgical grade silicon (m-Si), it is difficult to transfer laboratory results to industrial practice [18]. The commercial practice is largely empirical. Operators add their favorite metallic promoters in addition to the use of catalysts.

Catalyst - Many metal silicides are able to reduce the activation energy of the reaction between silicon with methyl chloride. Silicides in m-Si and their interaction with copper play a significant role in the reactivity of silicon in the Direct Process. The exposed silicon surface contains a layer of protective SiO_2 that blocks reaction, except at where dislocations in the silicon intersect the surface. Reactivity occurs as a result of transfer of elements through the silicide particles that are at the surface and penetrate the SiO_2 layer [22 - 24]. It is speculated that copper silicides, with the lowest melting points and heats of formation, contribute to higher reaction rates. Copper is known to have the highest diffusivity of any metal in silicon, and that it catalyzes the reaction at a relatively low threshold temperature in comparison with other metal compounds [20].

Copper forms 8 silicides. Despite the conflicting results in the literature, it is generally accepted that Cu_3Si represents the primary compound that facilitates the formation of Me_2SiCl_2 [25 - 27]. Cuprous chloride (CuCl) is the common precursor in the formation of Cu_3Si and responsible for the surface activity between Si and Cl [19]. Reaction between CuCl and m-Si appears to be dependent on the presence of particles of $CaSi_2$ and Al_2CaSi_2 in the silicon. Cuprous chloride and partially oxidized copper are the primary sources of copper used as catalyst in the Direct Process.

Partially oxidized copper consists of layers of $Cu/Cu_2O/CuO$. Both Cu and Cu_2O show high selectivity towards Me_2SiCl_2, while Cu or Cu_2O alone do not exhibit catalytic properties for this reaction [25]. Recently the significance of the surface area of the copper catalyst was demonstrated by preparing mesoporous Cu_2O microspheres [27]. The increased surface area, when compared with non-porous particles, led to a rise in the quantity of Cu_3Si formed, and consequently, stimulated the overall catalytic activity to produce Me_2SiCl_2. Moreover, selectivity towards Me_2SiCl_2, and Si reactivity increased when compared to when non-porous Cu_2O was used.

Promoters and Inhibitors – Small amounts of other elements are added to the copper catalyst. Those additions impact reactivity and selectivity through incorporation in silicides, and are referred to as either promoter or inhibitor depending on the impact the element, or admix of elements, has on the reaction. Promoters are thought to favor the diffusion of silicon atoms through silicide particles to where they react with chlorine on the surface of Cu_3Si particles [19 and 20]. Metal chlorides, for instance, show a strong impact on the reaction rate. Aluminum, cadmium, zinc and antimony chlorides have been shown to form active sites for reaction [28].

Zinc and tin with copper catalyst produced 90% selectivity for Me_2SiCl_2, and nearly 100% utilization of silicon [29]. Some promoters have a synergistic effect, an example being a Sn-Zn-Al combination has been found to achieve higher selectivity and reaction rate than the individual elements. It has been proposed that tin, as a promoter, with a low melting temperature and low surface tension enhances surface diffusion [30].

The concentration of tin in zinc affects the selectivity of Me_2SiCl_2. Increasing the tin content increases the production of $MeSiCl_3$, while diminishing that of Me_2SiCl_2 [20]. When zinc and tin are added as a mixture of separate elements, and not as an alloy, the mixture showed a beneficial effect on Me_2SiCl_2 selectivity [30 and 20]. Addition of aluminum along with tin helped to increase selectivity, whereas aluminum alone had a detrimental result [30].

Cesium, potassium, phosphorous and arsenic compounds have been suggested in a patent as promoters [31]. The inventors claim that replacing Sn with As in combination with Zn and P produces a promoter, whereas if Sn is not removed, the addition of As decreases selectivity. The addition of P and Sn combination was found to increase both reactivity and selectivity for Me_2SiCl_2 [32].

Cesium and rubidium are reported as promoters with m-Si, but act as inhibiters with pure silicon [33]. This result further supports the contention that a promoter alters the physico-chemical characteristics of silicides. When silicides are not present, the addition of a promoter element can only interact with the SiO_2 layer on the surface of the silicon. That layer is highly protective at the temperatures employed in the Direct Process.

Impact of Temperature and Pressure - Temperature and pressure are perhaps the easiest parameters to control in any industrial process. Variation of temperature and pressure in the Direct Process do not significantly impact the selectivity of products in a fixed-bed reactor running in the ranges of 1-26 bar and 553 to 633 K [34]. In another work selectivity of Me_2SiCl_2 produced in a fluidized-bed reactor was independent of temperature at 573 to 623 K [35]. In both cases, selectivity was related to the nature of the reactive mixture of catalyst, silicon and promoters, rather than to temperature. Similarly, in a separate investigation it was confirmed that temperature has only a small influence on product composition in the Direct Process, but has a positive effect on reaction rate [29].

Industrial Practice - Reaction of Si particulate with MeCl(g) (in the presence of a copper catalyst and promoters) in a fluidized bed reactor is carried out at temperatures between 543 K to 593 K and at total pressures of 1 to 5 bar [36]. In order to improve selectivity for Me_2SiCl_2 and reactivity, the silicon particulate must have a size between 1 and 85 microns [17]. That particle size range ensures a well fluidized bed, and thus good heat transfer in the reactor. Uncontrolled release of heat in the reactor can cause production of silicon tetrachloride and deposition of carbon [37]. Further efforts in enhancing reactor performance suggested the addition of inert materials, such as sand can improve temperature control and, hence, better control of the reaction rate [38].

Waste Products - Waste products from the Direct Process consist mainly of solid particles of Si, C, $CuCl_2$, CuCl, copper silicides, and other metal chlorides and silicides. This residue is highly unstable, with many of the chlorides and silicides readily reacting with atmospheric moisture to produce H_2(g) and/or HCl(g). Thus, the solid residue must be treated as a hazardous material. It must be deactivated before disposal.

T richlorosilane (TCS) and the Siemens Process

Hydrochlorination - In the semiconductor industry, chlorosilanes are common precursors for the deposition of silicon; therefore, high purity chlorosilanes are in high demand. The reaction of hydrogen chloride with m-Si at elevated temperatures, produces a variety of chlorosilanes. The products of the reaction include; chlorosilane (SiH_3Cl), dichlorosilane (SiH_2Cl_2), trichlorosilane ($SiHCl_3$), silicon tetrachloride ($SiCl_4$), and hydrogen as illustrated in the following generic reactions:

$$Si + (4-x)HCl(g) \rightarrow SiH_xCl_{4-x} + (2-x)H_2(g) \text{ (x = 0, 1, and 2)} \tag{2}$$

$$Si + yHCl(g) \rightarrow SiH_{4-y}Cl_y(g) + (y-2)H_2(g) \text{ (y = 2, 3 and 4)}. \tag{3}$$

The following discussion is focused on production of $SiHCl_3$ and its use in producing electronic-grade silicon (e-Si) in the Siemens Process. A flow diagram for that process is presented in Figure 2. $SiHCl_3$ formed in the hydrochlorination reactor is purified and then decomposed in the decomposition reactor (often referred to as the hairpin reactor as shown in Figure 3). The first report of reaction between silicon and hydrogen chloride gas to produce $SiHCl_3$ was published in 1857 [39]. One hundred fifty years later the typical distribution of chlorosilanes in the product stream is provided in Table 3. The reaction between hydrogen chloride and silicon to produce trichlorosilane is typically conducted at temperatures between 573 K to 623 K and at a pressure of 50 bar as represented by the reaction

$$Si + 3HCl(g) \xrightarrow{583K-633K} SiHCl_3(g) + H_2(g) \tag{4}$$

This reaction is highly exothermic ($\Delta H^o_{600K} = -218 kJ / mol\ of\ Si$) and as a result hot zones can form in a static bed reactor. As temperatures continue to rise, it has been shown that the percent yield of $SiHCl_3$ decreases, while the percent yield of $SiCl_4$ increases [39].

To address the exothermic nature of reaction 4, a variety of solutions have been proposed, including a fluidized bed reactor that is now standard practice. A number of patents for the production of $SiHCl_3$ in a fluidized bed reactor have addressed methods for achieving proper temperature control [40 and 41].

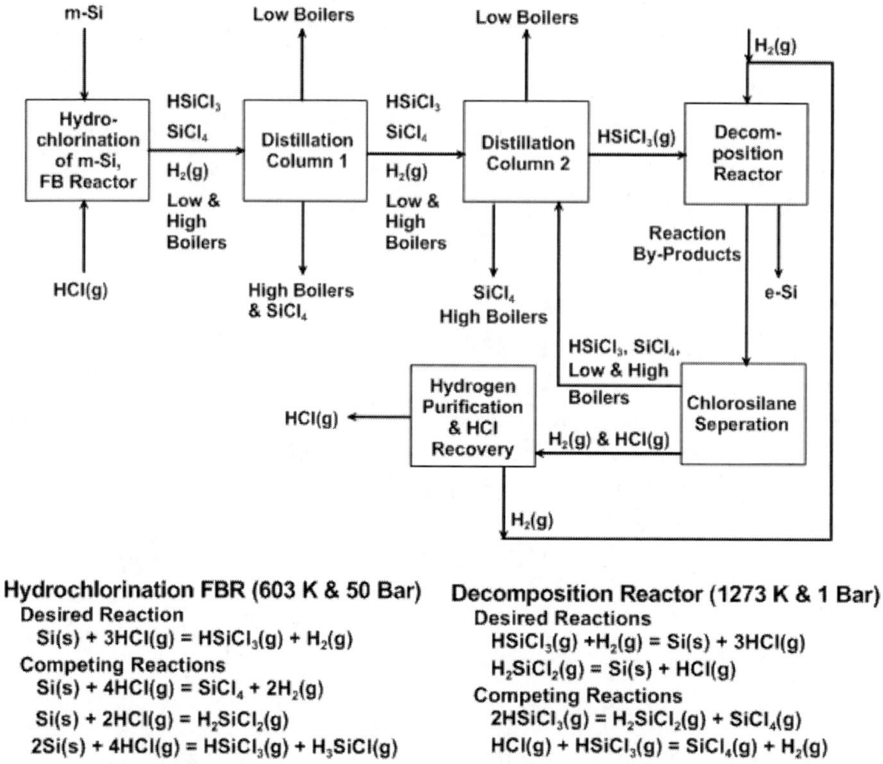

Figure 2. Flow Diagram for Siemens Process after ref.[41] (FB is fluidized bed).

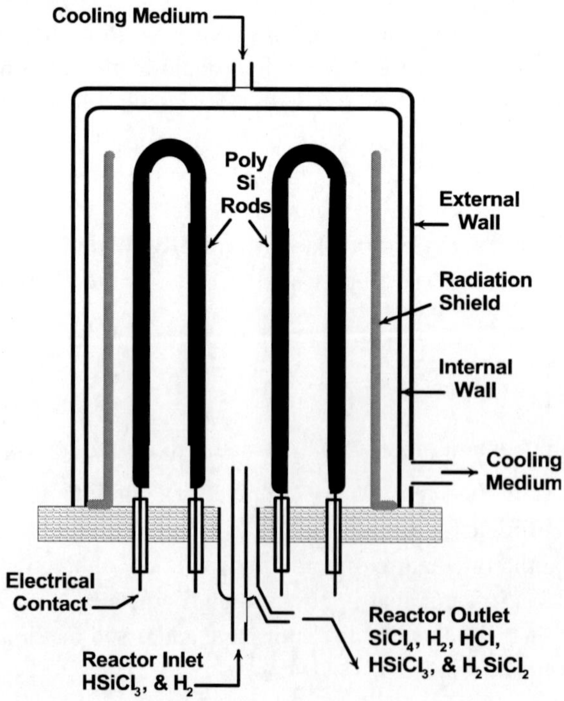

Figure 3. Hairpin Reactor for Decomposition of $SiHCl_3(g)$ [41].

air or air-N_2 purge at 298-503 K for 10-20 min.

: two steps are continuously repeated. It is clear that the yield of $SiHCl_3$ can be increased h the process, but there is also the added benefits of reducing unwanted by-products; $_3$Cl, SiH_2Cl_2, and most importantly $SiCl_4$. Since air can be used as an oxygen source, this thod would not require modification of the reactor. The net result is increased yield, at sumably decreased cost.

Siemens Process - $SiHCl_3$ is produced and decomposed in the Siemens Process. Figure 2 eals that m-Si is reacted with HCl(g) to product $SiHCl_3$, the resulting gas stream rich in Cl_3 is purified via distillation columns, and finally e-Si is produced by decomposition of purified $SiHCl_3$, where the silicon is deposited onto seed rods.

The Siemens Process is energy intensive, consuming 150 kWh per kg of Si produced at its with annual production of 5,000 metric tons. Over 90% of the energy loss is to the cold ls of the hairpin reactor [42]. The decomposition reaction of $SiHCl_3$ follows the following ction scheme

$$SiHCl_3(g) + H_2(g) \xrightarrow{1323K-1373K} Si + 3HCl(g) \tag{6}$$

s reaction is endothermic $\left(\Delta H^o_{1350K} = 211\, kJ/mol\ of\ Si\right)$ requiring a temperature of 3K to 1373K. Thus the heat requirement for the reaction, energy losses from the reactor ng both decomposition and during an extensive heat-up period, and the batch nature of process makes the operation of the decomposition reactor both energy intensive and ensive. In addition the conversion of $SiHCl_3$ is an unimpressive 25 to 35% with a single s. For every metric ton of purified silicon produced, 3 to 4 tons of silicon tetrachloride are ned. Yet roughly 70-75% of the world's polysilicon comes from the Siemens process [43]. vever, with the growth of the demand for silicon for photovoltaics overtaking that for on for the electronics industry in 2005, new energy efficient processes are erdevelopment or have recently come on line. An example of the latter is REC's new ess in operation in Moses Lake, Washington which is examined in a subsequent section his chapter.

Modifications to the Siemens Process - A variation of the Siemens Process involves cling products from both the hydrochlorination and the hairpin reactors [44] as shown in re 5. This should reduce cost by improving yield, and reducing the amount of $SiCl_4$ that t be disposed of.

The alternative process is centered around producing $SiHCl_3$ using a two-stage rochlorination reactor. The first stage involves the hydrogenation of $SiCl_4$ by the owing reaction:

$$3SiCl_4(g) + 2H_2(g) + Si \xrightarrow{773K-973K} 4SiHCl_3(g) \tag{7}$$

re, during operation of the reactor the ratio of H_2:$SiCl_4$ is maintained at roughly 2:1. A 2-eight percent of CuCl catalyst is mixed with the silicon particles, and the operating sure is roughly 3.5 bar [44]. The effluent from the first stage contains roughly 20%-23%

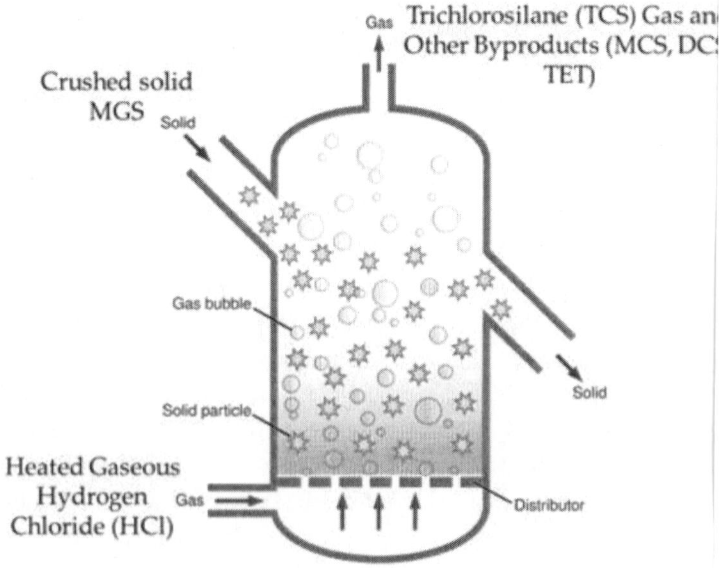

Figure 4. Hydrochlorination Fluidized Bed Reactor in Siemens Process.

Table 3. Comparison of product distribution from hydrochlorination r commercial operation to revised experimental procedure using inte oxidation [39]

Chlorosilane	Commercial Operation 1986 (% weight)	Experimental
SiH_3Cl	<0.1	-
SiH_2Cl_2	<0.5	~0
$SiHCl_3$	80-90	90-
$SiCl_4$	10-20	3-

Figure 4 is a pictorial representation of the reactor use to produce trichl been found that using metallurgical silicon of roughly 98% by weight s remainder composed of roughly 1% iron, 0.5% aluminum, and 0.1% calci higher yield of $SiHCl_3$ [39]. Silicon particles less than 300 microns yields bett ease of fluidization and better heat transfer [39].

It has been found, experimentally, that treating m-Si particles with c injecting hydrogen chloride gas into the fluidized bed reactor increases select [39]. Alternating the fluidization of silicon particles with air and then HC results presented in Table 3 where they are compared to commercial resul procedure for this process is as follows:

First

$$Si + 3HCl(g) \xrightarrow{573K-598K} SiHCl_3(g) + H_2(g) \text{ for 2-3 hours}$$

Second

$SiHCl_3(g)$. The second stage of the two-stage reactor is operated as the hydrochlorination reactor in the standard Siemens Process, namely reaction between m-Si and HCl(g). The product stream leaving the second stage contains 90% by weight $SiHCl_3$.

Union Carbide / REC Process

REC recently brought its FBR process online, a process originally pioneered by Union Carbide. After the oil crisis in the 70s, the U.S. Government funded several projects with the objective of finding an inexpensive route for producing solar silicon. Union Carbide achieved considerable success with the process presented in Figure 6. In 1990 Union Carbide sold the process to Komatsu, which became Advanced Silicon Materials (ASiMI) that was acquired by REC in 2005. Unique aspects of the process include using $SiCl_4$ as a feed material for reaction with m-Si in producing $SiHCl_3$, possibly using the 2-stage reactor discussed above.

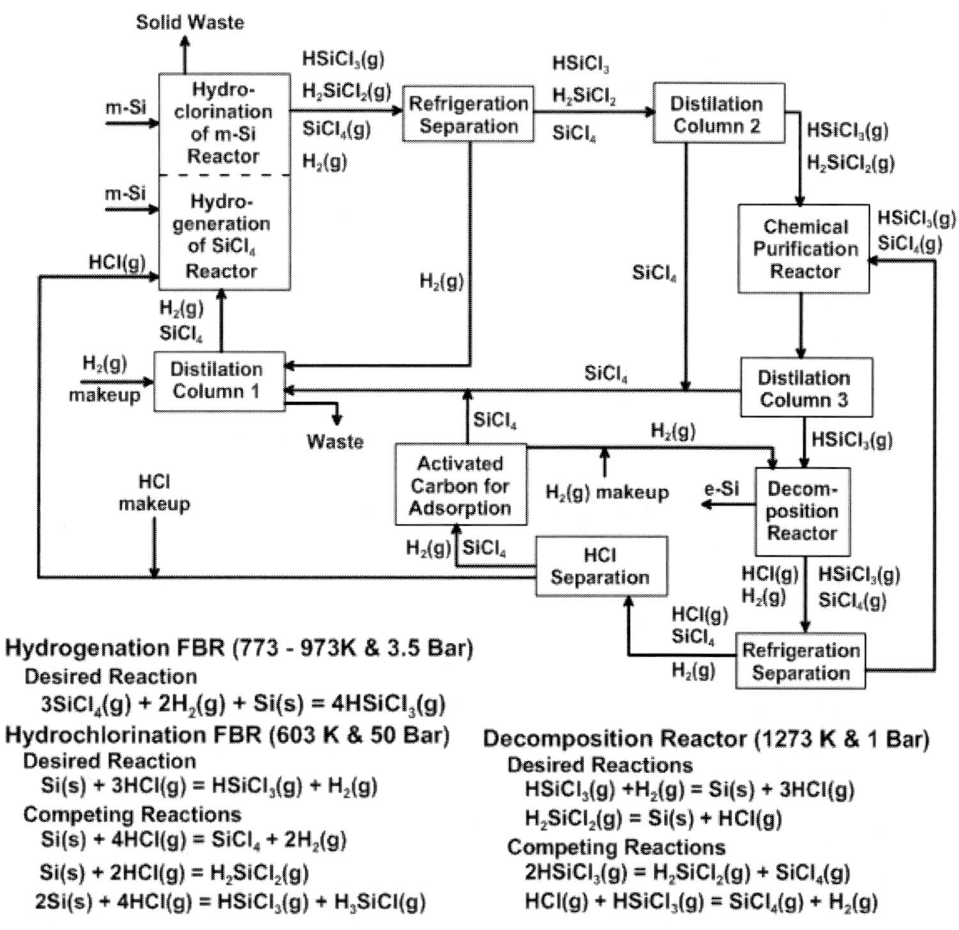

Figure 5. Revised Siemens Process showing recycle of byproducts to a 2-stage hydrochlorination reactor [44].

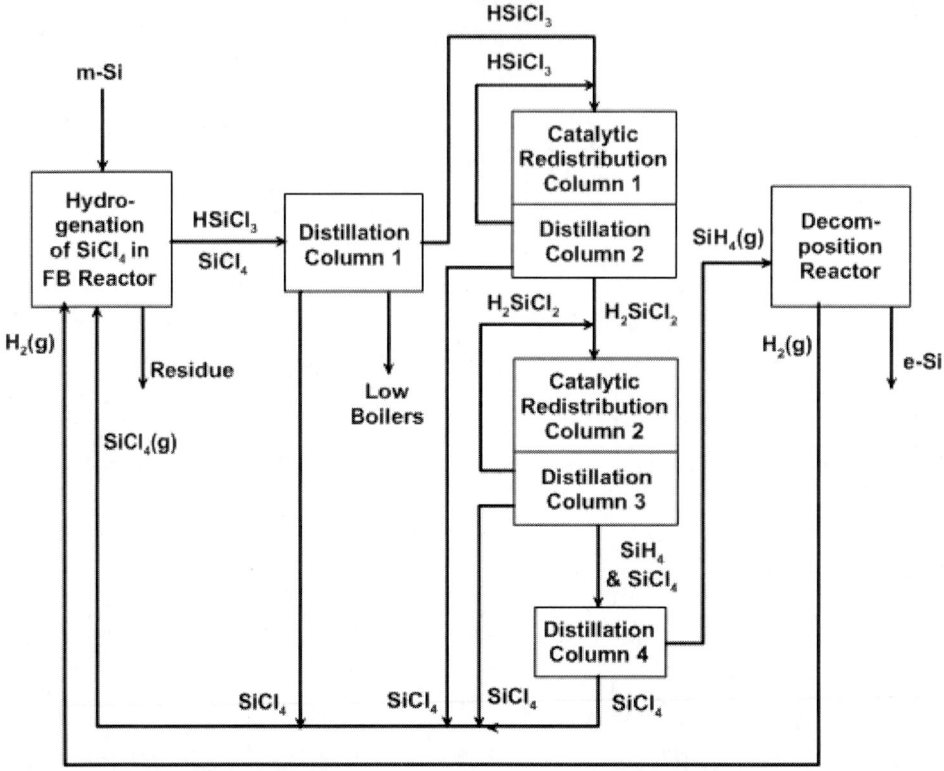

Hydrogenation FBR

Desired Reaction

$$3SiCl_4(g) + 2H_2(g) + Si(s) = 4HSiCl_3(g)$$

Decomposition Reactor

Desired Reaction

$$SiH_4(g) = 2H_2(g) + Si(s)$$

Catalytic Redistribution

Desired Reactions

$$2HSiCl_3 = H_2SiCl_2 + SiCl_4$$

$$3H_2SiCl_2 = SiH_4 + 2HSiCl_3$$

Figure 6. Flow Diagram for the Union Carbide Process after ref. [42].

The diagram in Figure 6 suggests that the process is self contained with respect to $SiCl_4$, however information provided by REC indicates that is not entirely the case. It appears that the company at minimum purchases some silicon tetrachloride as makeup for losses not shown in the flow diagram. It is possible that the actual process does not employ all the $SiCl_4$ recycle streams shown in the diagram. The $SiCl_4$ stream leaving the first distillation column likely contains a significant percentage of other high boilers and impurities, and thus may not be recycled. High purity silicon tetrachloride may come from a company producing silicon with the Siemens process.

The $SiHCl_3$ produced in the hydrogenation reactor is cleaned by passing it through a distillation column before it is redistributed in two separate reactors where the $SiHCl_3$ is passed through a fixed bed containing a quarternary ammonium ion exchange resins that act as a catalyst [42]. The SiH_4 containing gas leaving the second redistribution reactor is cleansed of its $SiCl_4$ in the final distillation column and then fed to the deposition reactor where it is decomposed. SiH_4 readily decomposes at temperatures as low as 773K, although actual temperature employed in the process is unknown.

REC has used the hairpin decomposition reactor in Figure 3, but with silane as the feed material in lieu of $SiHCl_3$. They have developed a fluidized bed reactor for the decomposition of SiH_4 that is projected to save between 80 to 90 percent of the energy used in the traditional Siemens process for silicon refining. Thermodynamic data indicate that the decomposition of SiH_4 is energetically favored at room temperature. REC's startup of their FBR for decomposition of SiH_4 was not without problems. The company states in their 2009 annual report that shortly after starting the reactor it had to be shut down for safety reason, but they were soon able to restart the reactor after corrective action was completed [45]. The company went on to report that they are increasing run time, and gradually improving yield and quality. From the latest REC annual report it appears these problems have been solved.

Schmid Process

The first purified silicon produced using the Schmid Process occurred June 2011 in Germany [43]. The process, as represented in Figure 7, has similarities to both the Siemens and the Union Carbide/REC processes. $SiHCl_3$ is produced by a combination hydrochlorination of silicon and hydrogenation of $SiCl_4$. The process, like the others, also produces SiH_2Cl_2, SiH_3Cl, and $SiCl_4$. The Schmid Process, like the Union Carbide/REC Process seeks to purify and then convert the $SiHCl_3$ to SiH_4 for decomposition. The latter process relies on combination of distillation and catalytic redistribution to achieve a satisfactory purity of the SiH_4 for decomposition. In the Schmid Process the conversion of $SiHCl_3$ to SiH_4 is achieved by disproportionation, that is essentially a redox reaction that also produces some $SiCl_4$ [43]. Although there is no indication of a distillation column after the disproportionation reactor in the Schmid Process, it is likely there must be at least one to achieve separation of SiH_4 from $SiCl_4$ and other high and low boilers. The $SiCl_4$ is recycled back to the hydrogenation reactor.

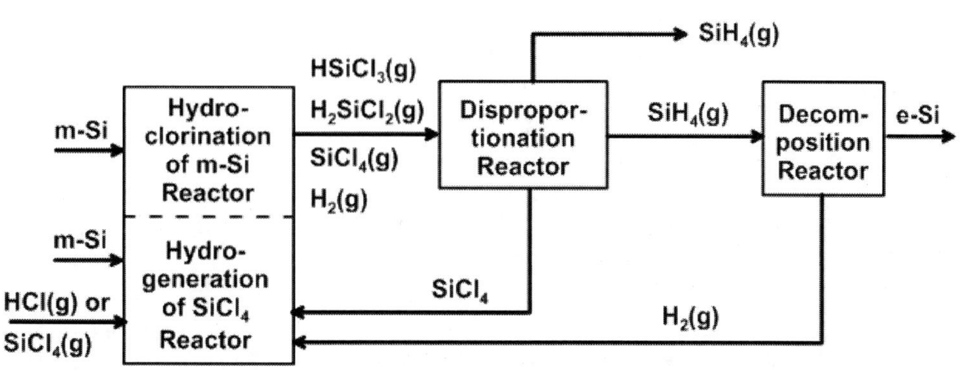

Figure. 7 Flow diagram for Schmid Process. The purified SiH_4 can either be decomposed to produce e-Si or sold as a separate product. [43].

The disproportionation of chlorosilanes is more efficient at higher temperature; however, operating temperatures are limited by the thermal stability of the catalyst, thus resulting in a typical operating temperature range of 278-285 K [43].

The keys to economic success in comparison of the Siemens, Union Carbide/REC, and Schmid processes is largely in the chemistry, design, and operation of the decomposition reactor. The SiH_4 product from the disproportionation reactor or the catalytic redistribution columns represents a valuable product that can be used to produce high purity silicon, or sold to the electronics industry for masking or formation of isolation layers (more on that later in this chapter).

Production of high purity SiH_4, that services two markets, provides an economic advantage for both the REC and the Schmid Processes. SiH_4 used to produce e-Si in the Schmid Process undergoes pyrolysis at roughly 923 – 1023 K in a hairpin reactor similar to that used in the Siemens Process [43].

The Schmid Process does have advantages over the Siemens process such as lower power consumption due in part to lower operating temperatures in the decomposition reactor (associated with decomposition of SiH_4 versus $SiHCl_3$), no need for byproduct disposal as $SiCl_4$ is recycled, and the availability of a variety of catalysts for disproportionation and the ability of some catalyst to remove As, P, and B thus yielding a higher purity e-Si [43].

REC uses a FBR to decompose the silane, a significant economic advantage not shared by the Schmid Process. The Schmid web site has a drawing suggesting they use a hairpin type reactor similar to that used in the Siemens process. Both have significant energy savings over the Siemens process, but REC's savings should be greater.

REC's decomposition reactor can be run continuously, where as the hairpin reactor is a batch operation. REC's reactor does not have an extensive heat-up time where there is little or no production, and no down time when the hairpin reactor must cool before removing the purified silicon.

Decomposition of SiH_4 on hairpins requires heating of the hairpins by passing electrical current through them. That requires getting the temperature in the reactor up to 900 K for the silicon to conduct the electrical charge before introducing silane. That will require two separate heating systems, since silicon acts as an electrical insulator at lower temperatures. Thus, capital cost for heating the hairpin reactor is significant greater, given the need for two electrical systems.

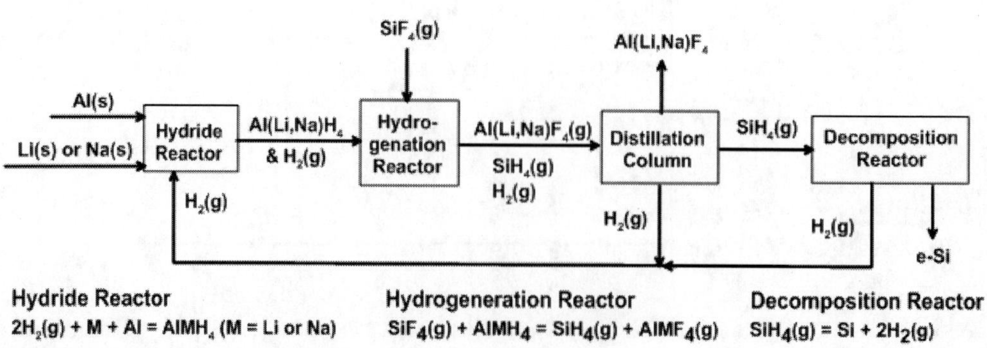

Figure 8. Flow diagram of Ethyl Corporation Process [42].

Ethyl / MEMC Process

The Ethyl Process, like the Union Carbide Process, got its start due to government concerns over availability of energy after the oil crisis in the 70s. The Ethyl Corporation, a long time producer of e-Si, teamed with the Monsanto Electronic Materials Company (MEMC) in 1987 to further advance years of research by building a plant in Pasadena, Texas for producing high purity granular silicon, originally intended to service the photovoltaic industry. In 1995 MEMC purchased the facility. The plant produces silicon for both the electronics and photovoltaics industries, as well as semiconductor-grade SiH_4.

While the Ethyl/MEMC Process produces a high purity stream of SiH_4 for decomposition, similar to that found in both the Union Carbide and Schmid processes, that is where similarities end. The flow diagram for the process presented in Figure 8 reveals that unlike the other processes, the Ethyl/MEMC Process does not rely on m-Si or $SiCl_4$ as the raw source for silicon, nor does it produce $SiCl_4$. Instead it uses SiF_4, a waste material from the manufacture of fertilizer, as the source of silicon.

The first step in the process is to produce a hydride reactant that can separate fluorine atoms bonded to silicon atoms, and in the process produce $SiH_4(g)$. The hydrides of Al and Li ($AlLiH_4$), and Al and Na ($AlNaH_4$) are both highly favored thermodynamically for exchanging the hydrogen atoms in the hydrides for fluorine atoms in SiF_4 at room temperature to temperatures over 1000K. The process is exothermic so cooling must be supplied. The reactions involving the formation of the hydrides are not as thermodynamically favorable, as hydrides as a class of materials tend to be unstable. At ambient pressure the maximum operational temperature is near 500K. Higher temperatures can be employed, but that would require higher total pressure of operation. The formation of the hydrides is also an exothermic reaction. A distillation column is used to separate $SiH_4(g)$ from the Al-metal fluorides and $H_2(g)$. The latter is recycled back to the hydride reactor, and the SiH_4 is decomposed to produce the high purity silicon.

Early on the Ethyl Corporation recognized the need to eliminate the hairpin reactor used in the Siemens Process in favor of a fluidized bed reactor(FBR) for decomposition of the $SiH_4(g)$. Their patent for decomposing silane in a FBR was issued in November of 1988 [46]. The commercial importance of the FBR versus the hairpin reactor are significant, as noted previously.

Other Methods for Producing Silanes

There are numerous methods for producing silanes that are not covered in this review due to time and space limitations. A few of those processes are listed in Table 4 and references are provided for those seeking additional information.

METHODS FOR PURIFICATION OF SILANES

Impurities - The semiconductor and photovoltaic industries require silicon to be of high purity; for the semiconductor industry, nine-nines pure (99.9999999%), while for the photovoltaic industry six-nines pure is acceptable. In order to produce high purity silicon, the SiH_4(g) and $SiHCl_3$(g) used in many processes to produce that silicon must also be of high purity. The impurities present in those gases depends on the manufacturing technique used.

Table 4. Alternative Methods for Producing Silanes

Product	Reactants	Catalyst or Promoter	Process Conditions	Ref.
$SiH_2(MgOH)_x$ SiH_4 and $M(OH)_x$	Metal Silicides (M_xSi) and H_2O $SiH_2(MgOH)_x$	HCl H_2O	673 - 773 K	[47]
R_nSiH_{4-n} and MCl	R_nSiCl_{4-n} (R is any radical including Cl) and Metal Hydride (MH)		298 K	[48]
$HSiCl_3$	$SiCl_4$, H_2 and Si		723 K, 34.5 bar	[49] and [50]
SiH_4 and $SiCl_4$	$HSiCl_3$	Halides of Al, B, Zn, and Fe	403 - 408 K	[51]
$HSiCl_3$, $SiCl_4$ and H_2	Si and HCl		573 - 698 K, 0.02 - 0.05 bar	[52]

As a result, there is a wide assortment of possible impurities that can be found in the silanes. Some common impurities present are, but not limited to, hydrogen, ammonia, helium, arsines (AsH_3), phosphines (PH_3), diboranes (B_2H_6), carbon monoxide, and simple and complex functional silanes. In order to remove these impurities, two common procedures are used to achieve a higher purity; distillation and adsorption.

Distillation - Distillation is a physical separation method where components in a mixture are separated based on differences in solution composition, temperature, and pressure required for boiling of those components. The separation of the components occurs because the vapor of a boiling mixture will have a higher concentration of the components that have lower boiling points, known as low boilers. The liquid residue retains the high boilers. It is possible to achieve separation of the components in the original mixture by slowly cooling the vapor over a physical distance such that specific components in the gas condense at different temperature and into separate trays. Depending on the impurity and its boiling point with respect to the boiling point of silane, the impurity, at the conclusion of distillation, can be in either the vapor phase or the liquid phase. For instance, distillation involving an impurity with a lower boiling temperature will produce a vapor with a high concentration of that impurity, which is then removed.

This method of removing impurities is a very simple approach but is also a very problematic one. Although many of the impurities commonly found in silanes have exploitable differences in boiling temperature, there are some impurities with very similar boiling temperatures as the $SiHCl_3$ and SiH_4. As a result, these impurities can't be removed by distillation. Another drawback of distillation is its high energy cost. A lot of energy goes into the heating and cooling of the mixture and all this energy increases the costs of producing purified silanes. Due to these disadvantages, other forms of purification have been developed.

Adsorption Techniques - Adsorption is defined as the adhesion of atoms or molecules to a surface. In terms of purification, adsorption is a process where a component in one phase is removed from that phase and accumulates on the surface of a different phase. While distillation used the differences in the boiling points between components, adsorption uses the differences in surface tension to cause separation. This difference in surface tension causes a component to be drawn out of one phase and adsorb onto another. Adsorption can be a completely physical process in which Van der Waals interactions produce the attraction so that no chemical properties are altered (physical adsorption). Adsorption can also be a process where the component forms a chemical bond to the surface that it has adsorbed onto (chemisorption). For the purification of silane gas, chemisorption is used instead of physical adsorption because it allows for more control over what component is adsorbed onto a surface.

Depending on which impurity is being targeted for removal, different solid surfaces may be used for this process. Some common materials used are activated carbon, silica gel, and zeolites. However, while the material used to make up the solid surface may vary, the materials themselves all have similar properties. In order to be a good adsorbent, the material must have a high surface area. A higher surface area provides more sites onto which impurities may adsorb onto. The pore structure of the material is also of importance. The pore size must not be smaller than the molecule size of the target impurity, or those impurity molecules will not be able to enter the pores.

Adsorption for purification is attractive because it allows for the removal of all impurities, whereas distillation is limited to only those impurities with significantly different boiling temperatures. However, adsorption does have many downsides as well. While distillation has a high cost due to its use of a lot of energy required to heat and cool mixtures, adsorption is also costly in that it requires careful production of adsorbents. The adsorbents must be tailored so that it has a specific pore size and composition in order to remove sufficient amounts of impurities. Furthermore, the adsorbents need periodic replacement or cleaning since the impurities will collect on the adsorbent during use.

Additional Purification – The purification requirement for e-Si is increasing from 9-nines to 11-nines. Use of multiple distillation columns and/or adsorption techniques can remove B and P to 1 ppbw, and As to less than 1 ppbw [53].

Higher purity silicon can be achieved by converting $SiHCl_3$ to SiH_4. That transformation also converts As, B, and P impurities to hydrides. Following hydride formation high-reflux distillation can remove all arsenic hydrides and most of the boron and phosphorus hydrides. Boron hydrides can be removed by means of chemical-absorption through use of treated solid resin for chemical-absorption. Phosphorus hydrides can be removed through adsorption by a phosphorus-specific mole sieve [53].

Applications of Silanes in Producing Integrated Circuits (IC)

Silanes, besides being used as precursors for producing e-Si, are also used to build the internal circuitry for micro- and nano- processors. Currently, nanoprocessors are comprised of more than ten layers of transistors and interconnects thus allowing for high densities of transistors per unit area. Figure 9 is an example of an older microprocessor and the complex network of interconnects. All of these interconnects must be insulated from each other. SiO_2 is a great insulator for the application, but must be either thermally grown, or grown by using silane based precursors in a CVD reactor. The latter is the preferred choice if high temperatures used to produce thermally grown oxide would cause damage.

Silanes can also be used to create silicon based gate architectures. Figure 10 shows how silicon oxides and silicon nitrides created by the decomposition of silanes in a CVD reactor are used to create key features of n- and p- type gates in IC devices.

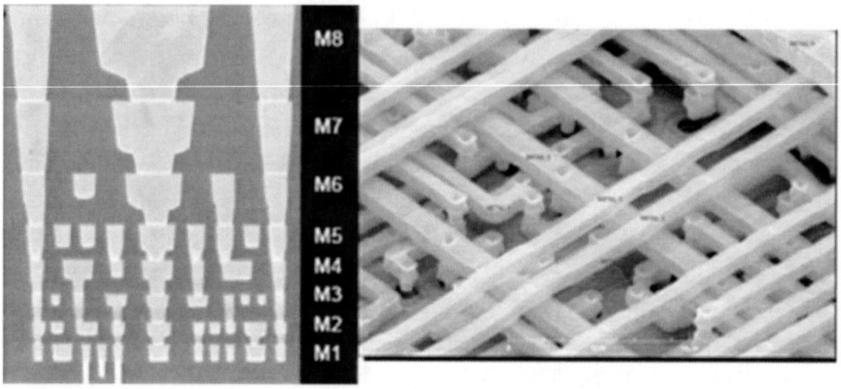

Figure 9. Image of the interconnect network in an IC device (right side). Left side of figure shows cross section of 8-layer device were light areas represent interconnects [54].

As previously mentioned, silanes can be used to grow oxide and nitride layers. In the presence of oxygen, oxygen will attack the silicon-hydrogen bond. When oxygen is in excess, the hydrogen will form a bond with oxygen thus producing silane radicals and water [56]. This typically occurs in atmospheric-pressure-chemical-vapor-deposition (APCVD) reactors. In low-pressure-chemical-vapor-deposition (LPCVD) reactors, oxygen is present is sparse quantities.

When oxygen in a LPCVD reactor attacks the silicon-hydrogen bond, silane radicals and hydrogen gas are produced [56]. Figure 11 shows a rough depiction of a proposed mechanism of how oxygen attacks silane. The SiH_4 molecule in the lower left of the figure can now bond with the silicon substrate, becoming part of the silicon surface. Additional condensation reactions such as those shown in the lower portion of Figure 11 can occur, thereby removing the remaining hydrogen from the surface and leaving behind a silicon oxide film.

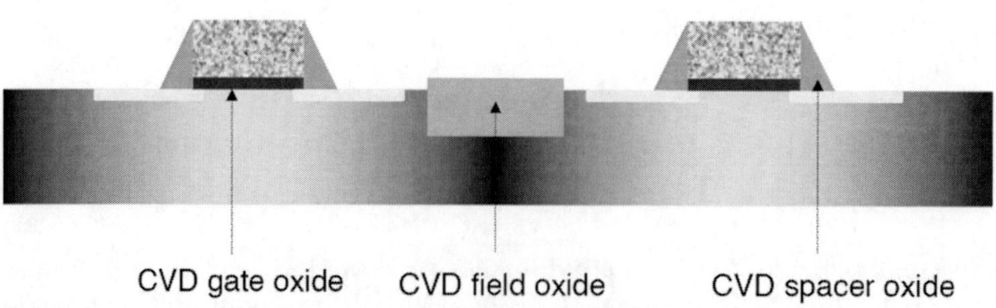

Figure 10. Silanes are decomposed along with other reactants to produce key feature in IC devices [55].

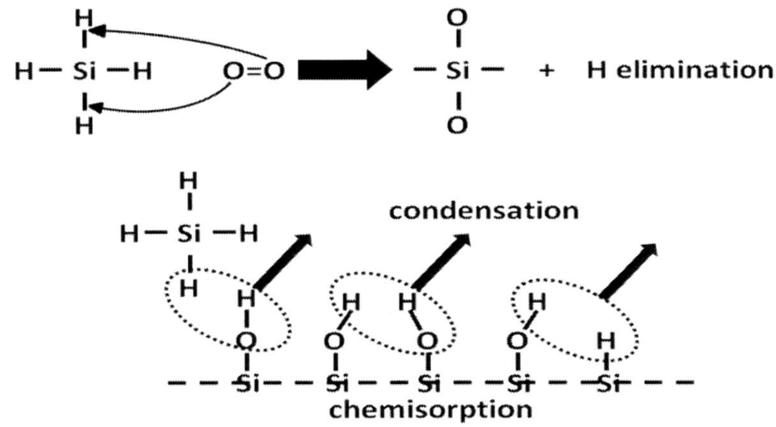

Figure 11. Molecular view of how silane interacts with oxygen in CVD processes. Upper diagram represents LPCVD, while the lower diagram depicts APCVD after [56].

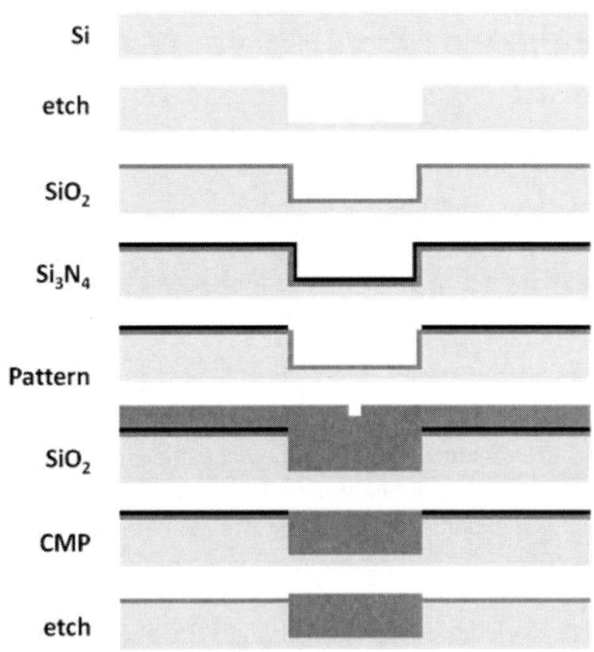

Figure 12. Sequence of events from top to bottom involving laying down layers and removing layers. SiO_2 and Si_3N_4 deposited through use of silanes (CMP is Chemical and Mechanical Polishing) [55].

Chlorosilanes are also used to grow features on an IC device. For instance, passivation layers containing silicon, oxygen, and nitrogen are grown by decomposing SiH_2Cl_2 in the presence of nitrous oxide and ammonia gas. Passivation layers provide electronic stability by isolating the transistor surface from electrical and chemical conditions in the environment.

Silicon nitride deposition is commonly used to provide a diffusion barrier and/or as an etch mask. Figure 12 gives a rough idea of how one might construct a given layer in an IC device by using the decomposition of silanes. Layers are added by CVD involving the use of silanes and layers are removed by either wet or dry etch. All of the steps that involved growing SiO_2 and Si_3N_4 used either silane or dichlorosilane as a precursor. Figure 13 shows the reaction schemes used to describe how insulators, isolators, and passivation layers are created by decomposing silanes in the presence of an oxidizer [55].

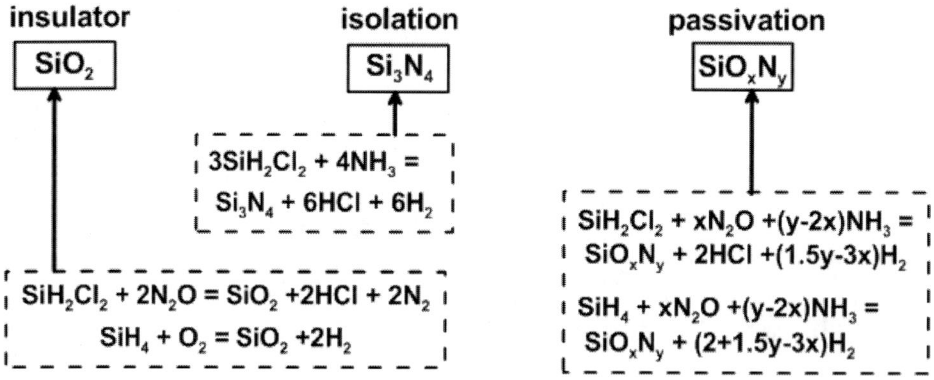

Figure 13. Silanes and the chemical reactions involved in producing layers in IC with desired results.

HEALTH AND SAFETY

Overview - SiH_4 is a dangerous and explosive gas posing numerous hazards in the occupational setting. This pyrophoric gas requires special equipment and handling precautions to manage it safely. SiH_4 has been responsible for numerous incidents including at least seven fatalities (U.S. and International) in addition to many injuries, and hundreds of thousands of dollars in economic losses. The majority of the sensational incidents involving silanes have occurred in specific industries including semiconductor manufacturing, photovoltaic cells, and specialty gas manufacturing. Several of these users have had serious accidents especially in the context of semiconductor applications [57]. A survey of twelve U.S. semiconductor manufacturers showed that there were 36 incidents between 1982 and 1997 [58]. Of the 36 incidents they included 15 fires in ducts and process tools, 6 fires from silane leaks in cabinets or gas supply systems, 5 explosions in ducts, and 3 explosions in cabinets or gas-supply systems. These accidents involved different parts of the SiH_4 system, from changing cylinders to emission control. Of these 36 reported incidents one serious injury (burns) and two minor ones (burns and temporary hearing loss) were reported. Forty-five percent of the silane incidents occurred during processing, 21% during maintenance and 21% during cylinder changes. Together, they caused about $500,000/yr in financial losses and business interruptions [58]. During the same period of time, three international incidents resulting in four fatalities were reported in Japan (see Table 5). The U.S. EPA Risk Management Program's (RMP's) national database includes six accidents that were associated with silane use in silane-gas manufacturing and at crystalline-silicon producer facilities from 1994 to 1999. Four incidents involved SiH_4 only, while the other two were attributed to SiH_2Cl_2 or $SiHCl_3$ leakage. None of these accidents resulted in deaths or severe damage, but two injuries and $686k in economic losses were reported [60]. In 2005 a routine procedure at a manufacturing plant in Taiwan caused a spontaneous explosion that killed a worker and ignited a blaze that ripped through the factory, shutting down production for three months. Another catastrophic event in 2007, and one of the most recent incidents involving a fatality, occurred outside Bangalore, India where an explosion decapitated an industrial worker, hurling his body through a brick wall [61]. On March 23, 2012, at a silicon chip manufacturer in Portland, Oregon two workers were injured while inside a vessel containing $SiHCl_3$, when oxygen leaked inside sparking a chemical explosion [62].

Table 5. Summary of Fatal Accidents in Japan between 1988 – 1998 [59]

Case	Year	Site	Deaths	Point of origin	Cause
explosion in silane cabinet	1989	Tokyo	1	valve vacuum pump	improper operation of valve
explosion of silane cylinder at university	1991	Osaka	2	check valve	failure of design, and degradation of rubber seal
explosion during change of silane cylinders	1996	Shizuoka	1	valve (closing nut)	inspection missed structure design flaw

Of all of the pyrophoric gases, SiH_4 is the most unpredictable despite 40 years of research. Silane has caused more fatalities and damage than all the other Electronic Specialty Gases (ESG) combined that include: arsine, phosphine, and hydrogen selenide in the past 20 years. The major hazards with SiH_4 are its pyrophoric nature, and pressure it must be stored at. SiH_4 will react in one of the following manners; no ignition, immediate ignition, or delayed ignition (explosion) [63]. There have been seven fatalities involving SiH_4 and gas cabinet explosions.

Hazards Research Gas Cabinet Testing - Hazards Research in New Jersey was able to simulate the 1976 fatality accident that occurred in Germany where they released pure $SiH_4(g)$ at two different points (at 20c m duct inlet, and at window level) into a 2 cylinder gas cabinet (44.5 cm x 91.4 cm x 216 cm) while it was being ventilated at 14 m^3/min. SiH_4 was discharged at pressures of 3.45 and 34.5 bar through 0.38, 1 and 4 mm orifices. The SiH_4 released through the 0.38mm orifice burned under all conditions. SiH_4 discharged through the 1mm orifice at 34.5 bar produced a violent detonation after a few seconds. The cabinet disappeared in 1/18 of a second, being blown apart with pieces flying over 30 m. The discharge at 3.45 bar was stopped after 16 seconds when it "popped", blowing open the cabinet doors. With the violent explosion the shockwave was well above 6,900 bar/s indicative of a very strong explosion [64].

Silane Leakage and Behavior - Cylinder connections for high purity gases are mechanically weak. Leaks have occurred in systems from under- or over- torque of joints, rapid cooling of connections due to the Joule-Thomson effect, and impact, mechanical stress, vibration, and twisting of connections.

Use of an elastomer gasket at high pressure, or cold flow, have produced leaks [63]. Small leaks of SiH_4(g) can produce the following; emissions without visual indication, popping sounds, solids formation, or fire with continuous flow or puffs. Whereas medium and large leaks may not immediately ignite, but explode with delayed ignition. Ignition may occur following abrupt shutoff, instances of unexplained ignition have occurred, and in some cases no ignition at all.

CODES AND STANDARDS

U.S. codes and standards for silanes have been enacted and include National Fire Protection Association (NFPA) 318 and 55, International Fire Code (IFC) chapters 18, 27, 30 and 40, American National Standard Institute's (ANSI) Compressed Gas Association (CGA) ANCI/CGA G–13. Key elements of CGA G-13, that are not in the other standards includes: comprehensive guidelines for bulk delivery systems, covers all applications (not just semi-conductor), and includes manufacturer filling facilities, provides comprehensive design guidelines, site plans, life safety, and gas properties [63]. *Health and Life Safety* - One of the primary tools used for identifying hazards associated with a material and its safe handling is the Material Safety Data Sheet, commonly known as the MSDS. Union Carbide has one of the earliest MSDS for SiH_4 dated 01-01-1970, while one of the most recent MSDS for SiH_4 comes from Dow Corning dated 09-19-2011 [65]. To date there are more than 1,700 MSDSs for Silane and its related compounds in the semiconductor and chemical manufacturing business [66]. MSDS's contain valuable information in a uniform and standardized format, although not a mandatory format. *United States Environmental Protection Agency (EPA)* - The EPA provides basic information on Acute Exposure Guideline Levels (AEGLs) as they pertain to human exposure to potentially toxic materials. AEGLs represent threshold exposure limits for the general public, and are applicable to emergency exposure periods ranging from 10 minutes to 8 hours. Three levels AEGL-1, AEGL-2, and AEGL-3 are developed for each of five exposure periods (10 and 30 minutes, 1 hour, 4 hours and 8 hours) and are distinguished by varying degrees of severity of toxic effects. Values are presented for SiH_4 in Table 6.

Table 6. EPA's Acute Exposure Guideline Levels for SiH_4 7803-62-5 (Interim 11/1/07) [67]

Exposure Level	ppm				
	10 min	30 min	60 min	4 hr	8 hr
AEGL 1	100	100	100	NR*	NR
AEGL 2	170	170	130	80	42
AEGL 3	300	300	270	170	80

*Not Rated.

AEGL-1 is the airborne concentration (expressed as parts per million or milligrams per cubic meter [ppm or mg/m^3] of a substance above which it is predicted that the general population, including susceptible individuals, could experience notable discomfort, irritation, or certain asymptomatic, non sensory effects. However, the effects are not disabling and are transient and reversible upon cessation of exposure. AEGL-2 is the airborne concentration of a substance above which it is predicted that the general population, including susceptible individuals, could experience irreversible or other serious, long-lasting adverse health effects or an impaired ability to escape. AEGL-3 is the airborne concentration of a substance above which it is predicted that the general population, including susceptible individuals, could experience life threatening health effects or death. Airborne concentrations below the AEGL-1 represent exposure levels that could produce mild and progressively increasing but transient and non-disabling odor, taste, and sensory irritation, or certain asymptomatic non-sensory effects. With increasing airborne concentrations above each AEGL, there is a progressive increase in the likelihood of occurrence and severity of effects described for each corresponding AEGL. Although the AEGL values represent threshold levels for the general public, including susceptible subpopulations; such as infants, children, the elderly, persons with asthma, and those with other illnesses, it is recognized that those individuals could experience the effects described at concentrations below the corresponding AEGL [66].

It should be noted that the greatest threat to human safety and primary health hazard associated with silane is the potential for severe thermal burns from contact with flames resulting from spontaneous ignition of this gas!

Acute Potential Health Effects - Routes of exposure to SiH_4 are indicated below.

- *Eye Contact*: Silane can be irritating to the eyes. Decomposition of Silane will result in the production of amorphous silicon dioxide. Eye contact with particulates of amorphous silicon dioxide can be irritating.
- *Ingestion*: Ingestion is not a likely route of industrial exposure.
- *Inhalation*: Inhalation in high concentrations of this gas can result in headache, nausea, dizziness and irritation of the upper respiratory tract. Silane can be irritating to the mucous membranes and the respiratory system. Severe over-exposure via inhalation may result in pulmonary edema and adverse kidney health effects that exist with inhalation of crystalline silicon dioxide. Overexposure to high concentrations may result in thermal burns due to the pyrophoric nature of the gas.
- *Skin Contact:* SiH_4 is not corrosive to the skin, but can be an irritant. Decomposition of Silane will result in the production of amorphous silicon dioxide. However, when in contact with water, SiH_4 may form silicic acid, which is corrosive to the skin.
- *Carcinogenicity:* Currently, silane has not been found to be carcinogenic.

Toxicological Information - "In toxicology, the median lethal dose, LD_{50} (abbreviation for "Lethal Dose, 50%), LC_{50} (Lethal Concentration, 50%) or LCt_{50} (Lethal Concentration and time) of a toxin, radiation or pathogen are the doses requird to kill half the members of a tested population after a specified test duration. In general LD_{50} figures are used as a guideline as an indicator of a substance's acute toxicity" [68].

Below are the median lethal doses for silane:

- LC_{50} (Inhalation): Rat LC_{50} 9600 ppm/4 hours
- LD_{50} (Oral): No data currently available
- LD_{50} (Dermal): No data currently available

Table 7. Fusion, Flash, and Boiling Point Temperatures [69 – 76]

Compound	Boiling point (K)	Melting point (K)	Flash point (K)
Non-functional silanes			
SiH_4	161	88	pyro.*
Si_2H_6	258.9	140.7	pyro.*
Si_3H_8	326	155.7	pyro.*
Si_4H_{10}	381.2	183.2	pyro.*
Si_5H_{12}	426.3	164.3	pyro?*
Si_6H_{14}	466.7	228.4	pyro?*
Methylsilanes			
$MeSiH_3$	215.5	116	<233
$Me_2Si\ H_2$	253.4	122.8	<233
Methyl-, Ethyl-, and Propyl-silanes			
Me_3SiH	282	137	226
$EtSiH_3$	259.3	93	216
Et_2SiH_2	329	138.6	252
Et_3SiH	381	116	281
$PrSiH_3$	293	-	239
Pr_2SiH_2	382	-	293
Pr_3SiH	446	-	316
Phenyl-silanes			
Ph_3SiH	425	317	349
Ph_2SiH_2	533	<293	383
$PhSiH_3$	393	207	289
Halogenated-silanes			
SiH_3F	175	-	-
SiH_3Cl	243	155	-
SiH_3Br	275	179	-
SiH_3I	319	216	-
SiH_2F_2	195.3	151	-
SiH_2Cl_2	281.3	151	227
SiH_2Br_2	343.7	203	-
SiH_2I_2	422.7	272	311
$SiHF_3$	178	142	-
$SiHCl_3$	305	146	239
$SiHBr_3$	385	200	-
$SiHI_3$	493	281	
Phenyl-chloro-silanes			
$PhSiHCl_2$	474.8	-	341
$PhSiH_2Cl$	314	-	327
Ph_2SiHCl	416	-	378
Methyl-halogen-silanes			
$MeSiH_2Cl$	282	-	286
$MeSiHCl_2$	314	-	244
Me_2SiHCl	309	-	248

Pyro = pyrophoric. pyro?* = likely pyrophoric.

Thermal-Chemical Data for Silanes

Working with silanes requires special care. The numerous industrial accidents have their origin in improper use and handling of silanes. Proper handling of silanes begins with knowledge of thermo-chemical data of which the boiling and flash point temperatures are presented in Table 7.

When it comes to calculating heat-loads for processes, values for specific heats, and heats of formation are required. Values for the latter are presented in Table 8, while values for specific heats are presented in Table 9.

Equations for computing the standard state Gibbs energy of formation of silanes from their elements (Si, C, H_2(g), Cl_2(g), F_2(g), Br_2(g), and I_2(g)) are presented in Table 10. The data in that table for chlorosilanes has been plotted in the customary Ellingham-Richardson format in Figures 14 and 15 (except for the fact that the product is a gas at a pressure of 1 bar) as an example of what information can be extracted from the thermodynamic data [98].

Table 8. Heat of Formation and Entropy Values of Silanes [77-96]

Compound	ΔH_f^o (kJ/mol)	S^o (J/mol K)
Non-functional silanes		
SiH_4(g)	34.3 ± 2.0	204.65
Si_2H_6(g)	80.3	272.66
Si_3H_8(g)	121	347
Si_4H_{10}(g)	160	411.7
Si_5H_{12}(g)	199	476
Si_6H_{14}(g)	217.1-239.2	-
Methylsilanes		
$MeSiH_3$(g)	-29.1 ± 4.0	257
$Me_2Si\ H_2$(g)	-94.7 ± 4.0	-
Me_3SiH(g)	-163.4 ± 4.0	-
Ethyl-, and Propyl-silanes		
$EtSiH_3$(g)	-88	
Et_2SiH_2(g)	-141.5 ± 21	
Et_3SiH(l & g)	-254 ± 4 -217.5 ± 5	
Pr_3SiH(l & g)	-319 ± 5 -280 ± 6	
Methyl-chloro-silanes		
$MeSiH_2Cl$(g)	-210 ± 7	
$MeSiHCl_2$(g)	-393 ± 9	
Me_2SiHCl(g)	-282 ± 7	
Halide-silanes		
$SiHF_3$(g)	-1200.81	277.26
$SiHCl_3$(g)	-496.22	313.71
$SiHBr_3$(g)	-302.92	348.04
$SiHI_3$(g)	-74.48	375.01
SiH_2F_2(g)	-790.78	262.12
SiH_2Cl_2(g)	-320.49	286.72

Table 8. (Continued)

Compound	ΔH_f^o (kJ/mol)	S^o (J/mol K)
$SiH_2Br_2(g)$	-190.37	310.04
$SiH_2I_2(g)$	-38.07	326.79
$SiH_3F(g)$	-355 ± 2 -376.56	238.39
$SiH_3Cl(g)$	-136 ± 10 -141.84± 8	250.76
$SiH_3Br(g)$	-64 ± 9 -78.24	262.46
$SiH_3I(g)$	-2 ± 8 -2.09	271.02

Table 9. Specific Heats for Silanes [97] (J/mol)

Silane	T (K)		
	298	500	1000
	$C_{p\ (J/mol)}$		
$SiH_4(g)$	30.1	30.5	34.0
$Si_2H_6(g)$	79.0	106.9	144.6
Halogenated Silanes			
$SiHCl_3(g)$	75.5	88.9	100.1
$SiH_2Cl_2(g)$	62.1	78.0	94.8
$SiH_3Cl(g)$	49.9	70.3	88.1
$SiHF_3(g)$	63.5	81.0	97.4
$SiH_2F_2(g)$	54.2	73.0	92.5
$SiH_3F(g)$	47.2	65.3	88.6
$SiHBr_3(g)$	80.4	85.1	101.0
$SiH_2Br_2(g)$	65.5	80.0	95.2
$SiH_3Br(g)$	52.6	69.0	89.8
$SiHI_3(g)$	83.9	93.2	101.7
$SiH_2I_2(g)$	67.9	81.3	95.7
$SiH_3I(g)$	54.4	70.1	89.8
Methyl- and Methyl & Phenyl Silanes			
$Me_2SiH_2(g)$	91.6	128.9	184.4
$MeSiHCl_2(g)$	95.9	114.5	160.3
$PhSiHCl_2(g)$	109.7	208.2	-

Table 10. Standard State Gibbs Energy of formation of Silanes [97]

Compound	$\Delta G_f^o\ (kJ/mol) =$	Temperature Range (K)
$SiH_4(g)$	$29.40 + 8.750 \cdot 10^{-2} T + 4.749 \cdot 10^{-6} T^2$	300 - 1600
$Si_2H_6(g)$	$73.67 + 0.1740 \cdot T + 4.329 \cdot 10^{-6} \cdot T^2$	300 - 1600
$SiH_3Cl(g)$	$-145.3 + 8.435 \cdot 10^{-2} T + 3.436 \cdot 10^{-6} \cdot T^2$	300 - 1600
$SiH_2Cl_2(g)$	$-323.3 + 9.316 \cdot 10^{-2} T + 1.478 \cdot 10^{-6} T^2$	300 – 1600
$SiHCl_3(g)$	$-497.5 + 0.1091T - 4.803 \cdot 10^{-7} T^2$	300 - 1600
$SiH_3F(g)$	$-380.9 + 8.856 \cdot 10^{-2} T + 3.897 \cdot 10^{-6} T^2$	300 - 1600
$SiH_2F_2(g)$	$-794.2 + 9.872 \cdot 10^{-2} T + 2.768 \cdot 10^{-6} T^2$	300 - 1600
$SiHF_3(g)$	$-1203 + 0.1172T + 1.249 \cdot 10^{-6} T^2$	300 - 1600
$SiH_3Br(g)$	$-98.39 + 8.660 \cdot 10^{-2} T + 2.332 \cdot 10^{-6} T^2$	350 - 1600
$SiH_2Br_2(g)$	$-224.4 + 9.267 \cdot 10^{-2} T + 8.450 \cdot 10^{-7} T^2$	350 - 1600
$SiHBr_3(g)$	$-350.5 + 0.1081T - 9.349 \cdot 10^{-7} T^2$	350 - 1600
$SiH_3I(g)$	$-39.07 + 8.745 \cdot 10^{-2} T + 1.252 \cdot 10^{-6} T^2$	475 - 1600
$SiH_2I_2(g)$	$-104.0 + 9.200 \cdot 10^{-2} T + 1.743 \cdot 10^{-8} T^2$	475 - 1600
$SiHI_3(g)$	$-169.1 + 0.1033T - 1.243 \cdot 10^{-6} T^2$	475 - 1600
$MeSiHCl_2(g)$	$-427.8 + 0.1721T + 1.807 \cdot 10^{-5} T^2$	300 - 700
$Me_2SiH_2(g)$	$-79.52 + 0.2702T + 1.114 \cdot 10^{-5} T^2$	300 - 1500
$PhSiHCl_2(g)$	$-267.9 + 0.2723T + 2.566 \cdot 10^{-5} T^2$	300 - 800

The standard state Gibbs energy of formation of the chlorosilanes in the figures are plotted according to the following chemical reactions.

<u>$SiH_3Cl(g)$</u>

$2/3Si + H_2(g) + 1/3Cl_2(g\ at\ 1\ bar) = 2/3SiH_3Cl(g\ at\ 1\ bar)$ (8)

$2Si + 3H_2(g\ at\ 1bar) + Cl_2(g) = 2SiH_3Cl(g\ at\ 1\ bar)$ (9)

<u>$SiH_2Cl_2(g)$</u>

$Si + H_2(g) + Cl_2(g) = SiH_2Cl_2(g\ at\ 1\ bar)$ (10)

<u>$SiHCl_3(g)$</u>

$2Si + H_2(g) + 3Cl_2(g\ at\ 1\ bar) = 2SiHCl_3(g\ at\ 1\ bar)$ (11)

$2/3Si + 1/3H_2(g\ at\ 1\ bar) + Cl_2(g) = 2SiH_3Cl(g\ at\ 1\ bar)$ (12)

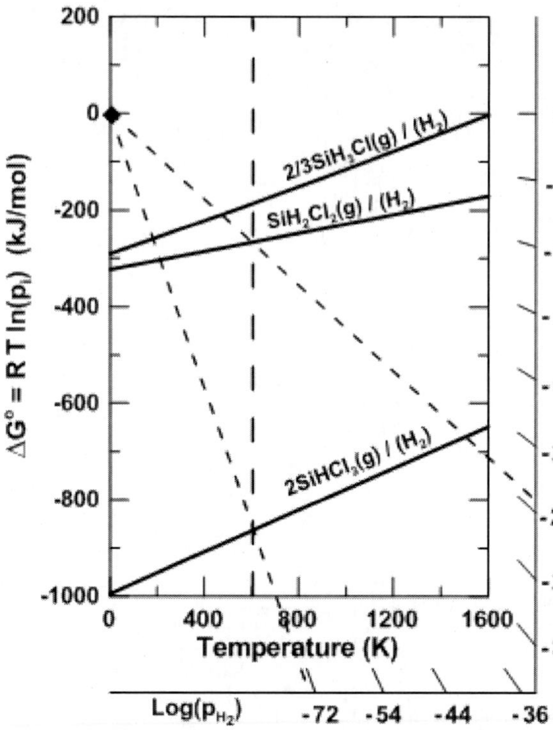

Figure 14. Ellingham-Richardson Diagram Used to Demonstrate the Importance of H_2 in Producing the Desired Product.

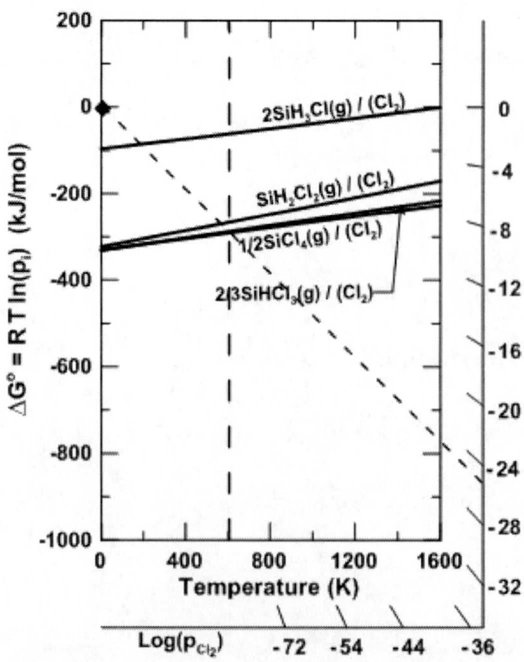

Figure 15. Ellingham-Richardson Diagram Used to Demonstrate the Impact of Cl_2 on Product Distribution in Producing Chlorosilanes.

The reactions have been written with respect to 1 mole of $H_2(g)$ or $Cl_2(g)$ so as to compare the stability of the silanes with respect to either gas. Only one reaction is written for SiH_2Cl_2 as the single reaction satisfies the requirement for 1 mole of either gas. In each figure the coefficient in front of the silane product is associated with the stoichiometry of the formation reaction. Following that designation is either a /H_2 or /Cl_2 or both to indicate that the scale to the right of the diagram pertains to the gas listed, and that the gas not listed is taken as having a pressure of 1 bar.

These stipulations limit the quantified utility of the diagrams to visualization of the equations in Table 10, but do provide significant qualitative information. Qualitatively the diagrams in Figures 14 and 15 deconstruct the impact of H_2 and Cl_2 on the thermodynamic equilibrium for the formation of the chlorosilanes. The impact of H_2 on the stability of the chlorosilanes in the presence of excess Cl_2 (at 1 bar) is presented in Figure 14. The diagram reveals that at 600 K $SiH_2Cl_2(g)$ at 1 bar would require a partial pressure of H_2 equal to approximately $5 \cdot 10^{-24}$ bar, where as only a value of 10^{-72} bar is required for formation of $SiHCl_3(g)$. Thus, from the diagram and a purely thermodynamic viewpoint, producing $SiHCl_3$ should be done with a reactant that produces a hydrogen potential (p_{H2}) significantly less than $5 \cdot 10^{-24}$ bar to avoid forming SiH_2Cl_2. Operating anywhere near that potential would produce significant quantities for both $SiH_2Cl_2(g)$ and $SiHcl_3(g)$ as the lines in the figure are for those gases at 1 bar. Thus, there are an infinite series of lines representing the reaction, but at different pressures for the chlorosilanes. At a lower pressure for the silane, the line is rotated clockwise and thereby decreasing the equilibrium pressure H_2.

The diagram in Figure 15 indicates, again from a purely thermodynamic viewpoint, that excess chlorine will lead to significant formation of $SiCl_4(g)$. To keep from making significant amount of the tetrachloride the pressure of Cl_2 must be significantly below 10^{-25} bar for the specified conditions. The problem one faces in producing the desired chlorosilane is to find a reactant that will produce the right combination of chlorine and hydrogen potential that will lead to an acceptable yield for the desired product at a reasonable production rate. The answer for formation of $SiHCl_3(g)$ has been to operate at low enough temperature where chemical kinetics dominate over thermodynamic expectations. Thus it has been possible to produce $SiHCl_3$ with more than 90% yield at temperature between 573K to 723K.

Decomposition of $SiHCl_3(g)$ to produce high purity Si in the Siemens process is accomplished at temperatures between 1323 K to 1423 K where thermodynamics plays a greater role. The diagram in Figure 15 indicates that at those temperature, and in the presence of excess $H_2(g)$ at 1 bar, there is a cluster of lines for $SiCl_4(g)$, $SiHCl_3(g)$, and $SiH_2Cl_2(g)$. Decomposition of $SiHCl_3(g)$ must include significant formation of $SiCl_4$ and SiH_2Cl_2. Industrial practice reveals that 3 to 4 tons of $SiCl_4$ is produced for every ton of silicon deposited. Formation of the tetrachloride is essential to reducing the pressure Cl_2 in the gas stream to keep the decomposition reaction going, but is accomplished at a loss of yield.

CONCLUSION

Organic silanes have found use as coupling agents, adhesion promoters, cross-linkers, water scavengers, and stabilizers. Functional silanes can be altered through addition of radicals to be either hydrophobic or oleophobic or both. The latter functional silane can act as a physical link between incompatible phases. The market for these silanes are expected to grow, but will mirror the world's economy.

The next great growth area for silanes will be in the production of silicon for photovoltaics. Chloro-inorganic silanes have been used in producing e-Si for more than 50 years. This market on a tonnage basis for silanes is small in comparison to that for the organosilanes, but one now poised to grow significantly. In 2005 more purified silicon was used in producing photovoltaics than that used in electronics devices.

The market for photovoltaics has grown phenomenally. In the last 19 years the annual installed capacity has grown over 30% over total installed capacity, except for 4 years in the 90s. In 2008, 2009, and 2010 the growth rate was almost 80%, over 40%, and over 70% respectively. Most of that growth has involved polycrystalline silicon (p-Si) as shown in Figure 16. Most of that growth was created with purified silicon produced in the Siemens Process. Cost reductions in producing purified silicon, established history for the use of silicon in photovoltaics, and the readily available source of raw material (SiO_2) ensure that silicon use in photovoltaics will be with us long into the future.

Siemens silicon is expensive with production cost at approximately $30 per kg. Significant cost reduction has been achieved in both the Union Carbide/REC and Schmid processes.

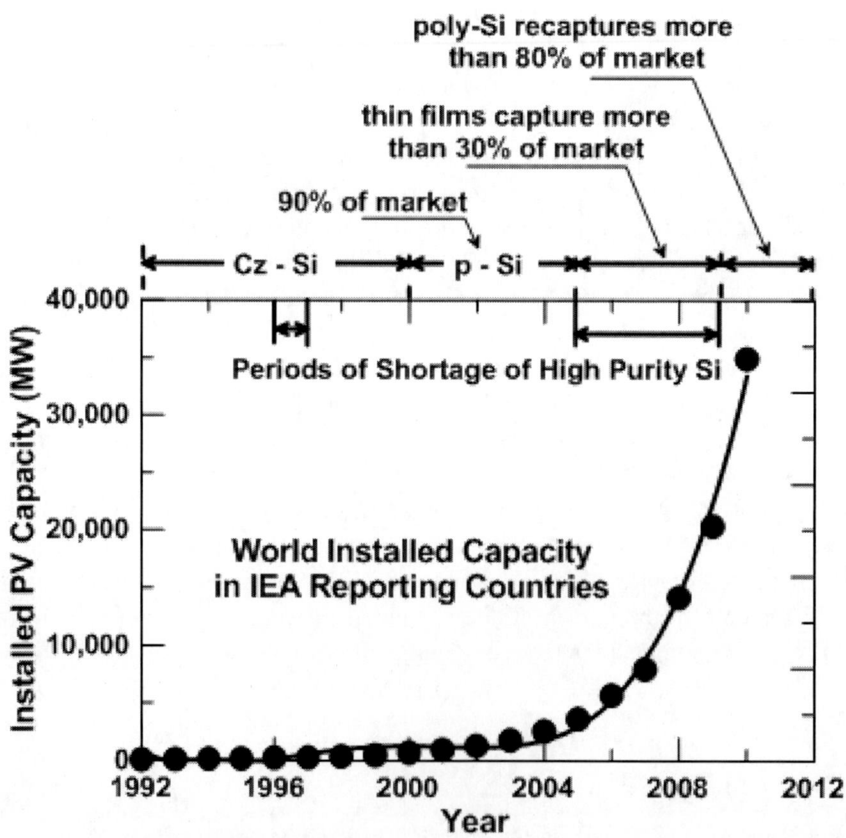

Figure 16. Growth in installed photovoltaic (PV) capacity and importance of polysilicon.

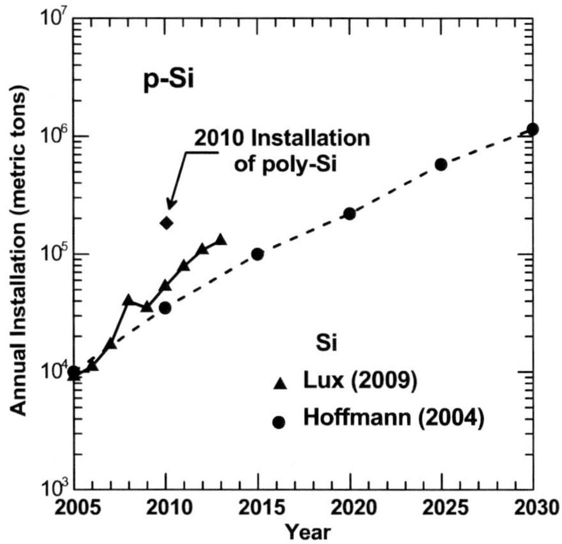

Figure 17. Estimates of consumption for polysilicon in photovoltaics, and comparison of actual figure for 2010 [99 and 100].

REC is the leader in production cost reduction at $15 per kg, and has expectations of reducing cost to below $12 per kg. REC and Schmid, like the Siemens Process rely on the use of silanes, but unlike the Siemens Process have substantially greater yield in converting SiH_4 to Si than that achieved with $SiHCl_3$. Actual usage of silicon in photovoltaics has far out stripped projections as shown in Figure 17. In the future there will be significant need for increased capacity for producing both $SiHCl_3$ and SiH_4.

The established use of organosilanes in consumer products and the defense industry, and the growth of the market for use of silanes as a precursor for producing purified silicon for both the electronics and photovoltaic industries insure that markets for silanes will continue to grow.

References

[1] Dow Corning. Silane Solutions: Full Range of Silanes [www page]. www.dowcorning.com/content/silanes/fullrangeofsilanes.asp (accessed 02/29/12).

[2] *Encyclopedia of Chemistry*, 4th ed., D. M. Considine, ed., Van Nostrand Reinhold Co., 1984, pp. 851-855.

[3] Dow Corning. Pharmaceutical Manufacturing [www page]. http://www.Dowcorning.com/content/silanes/silanespharma/default.asp (accessed 03/18/12), part of Dow Corning. http://www.Dowcorning.com/(accessed 03/18/12).

[4] Beach, H. L., Jr. et al. http://naca.larc.nasa.gov/search.jsp?R=19810007771andqs=Ns%3DNASA-Center|0%26N%3D4294824412 (accessed 02/19/12), part of NASA Technical Reports Server (NTRS). http://naca.larc.nasa.gov/search.jsp (accessed 02/19/12).

[5] Dow Corning. A Guide to Silane Solutions: The Basics of Silane Chemistry. https://www.xiameter.com/en/ExploreSilicones/Documents/Silane%20Chemistry-1a-95-718-01-F2.pdf (accessed 03/18/12), part of Silicones Simplified Xiameter from Dow Corning. https://www.xiameter.com/en/Pages/LandingPage.aspx (accessed 04/05/12).

[6] De Buyl, F. 19. Organo-Functional Silanes. http://www.Dowcorning. com/content/ publishedlit/chapter19.pdf (accessed 03/18/12), part of Dow Corning. http://www. Dowcorning.com/(accessed 03/18/12).

[7] Dow Corning. A Guide to Silane Solutions From Dow Corning. http://www.talleres norte.com.ar/pdf/pinturas/silanos/silane_guide.pdf (accessed 04/05/12), part of Dow Corning. http://www.Dowcorning.com/(accessed 04/05/12).

[8] Dow Corning. Mineral Fillers for Rubber and Plastics: An Introduction. http://www.rtvanderbilt.com/fillersintroweb.pdf (accessed 03/18/12), part of R. T. Vanderbilt Company, Inc. Industrial Minerals and Chemicals. http://www.rtvanderbilt. com/ (accessed 03/18/12).

[9] Dow Corning. Sealants and Adhesives. http://www.Dowcorning.com/content/ silanes/silanesseal/default.asp (accessed 03/18/12), part of Dow Corning. http://www. Dowcorning.com/ (accessed 03/18/12).

[10] Dow Corning. Paint – Ink – Coatings. http://www.Dowcorning.com/content/silanes/ silanespaint/ default.asp (accessed 03/18/12), part of Dow Corning. http://www. Dowcorning.com/ (accessed 03/18/12).

[11] Dow Corning. Silane Solutions: Existing Silanes Markets. http://www.Dowcorning. com/content/silanes/silanes_markets.asp (accessed 03/18/12), part of Dow Corning. http://www.Dowcorning.com/ (accessed 03/18/12).

[12] Dow Corning. Water Repellents and Surface Protection. http://www.Dowcorning.com/ content/silanes/silanessurf/default.asp(accessed 03/18/12), part of Dow Corning. http://www.Dowcorning.com/ (accessed 03/18/12).

[13] Evonic. Surface Modification. http://www.dynasylan. com/product/dynasylan/en/ effects/surface-modification/Pages/default.aspx (accessed 03/18/12), part of Dynasylan – Evonik's Brand for Functional Silanes. http://www.dynasylan.com/product/ dynasylan/en/Pages/default.aspx (accessed 03/18/12).

[14] Petrov, A. D.; Mironov, B. F.; Ponomarenko, V. A.; Chernyshev, E. A. Synthesis of Organosilicon Monomers; Consultants Bureau: New York, NY, 1964.

[15] Arkles, B.; Peterson, W. R., Jr. Silicon Compounds, Silanes. Kirk-Othmer Encyclopedia of Chemical Technology; 3rd ed; John Wiley and Sons, Inc, 1982; Vol. 20, p 887.

[16] Rochow, E. G. US Patent 2,380,995, September 26, 1941.

[17] Freeburne, S. K.; Halm, R. L.; Kohane, J. P.; Wineland, J. D. US Patent 5,312,948, May 17, 1994.

[18] Kalchauer, W.; Pachaly, B. in *Handbook of Heterogeneous Catalysis*; Ertl, G.; 2nd ed.; Wiley-VCH Verlag GmbH and Co, 2008; pp 2635-2647.

[19] Clarke, M. P. *J. Organomet. Chem.* 1989; Vol. 376, pp 165.

[20] Rethwisch, D. G.; Olakangil, J. in *Silicon for the Chemical Industry* VIII; Øye, H.; Brekken, H.; Foosnaes, T.; Nygaard, L. Norwegian University of Science and Technology. Trondheim, Norway, 2006; pp 103-120.

[21] Schei, A.; Tuset, J.; Tveit, H. *Production of High Silicon Alloys*. Tapir. Trondheim: Norway, 1998; pp 310-311.

[22] Halvorsen, S. A.; Schüssler, G. Norwegian Patent 169831 and German Patent 4037021, Nov., 1989 and May 1988.

[23] Rong, H. M.; Øye, H. in *Silicon for the Chemical Industry*; Rong, H. M.; H. Øye H. Institute of Inogranic Chemistry, NTH, Trondheim, Norway, 1992; pp 67-83.

[24] Rong, H. M. Silicon for the Direct Process to Methylchlorosilanes (Ph. D. dissertation); Norwegian University of Science and Technology, Dept. of Inorganic Chemistry, Trondheim, Norway, 1992.

[25] Floquet, N.; Yilmaz, S.; Falconer, J. L. *J. Catal.* 1994, Vol. 148, pp 348-368.

[26] Luo, W.; Wang, G.; Wang, J. Organosilane Synthesis. *Chem. Eng. Comm.* 2006; vol. 193, pp 754-763.

[27] Zhang, Z.; Che, H.; Wang, Y.; Gao, J.; Zhao, L.; She, X.; Sun, J.; Gunawan, P.; Zhong, Z.; Su, F. *Ind. Eng. Chem. Res.* 2012; vol. 51, pp 1264-1274.

[28] De Cooker, M. G. R. T.; De Jong, J. W.; Van den Berg, P. J. *J. Organomet. Chem.* 1975; vol. 86, pp 175-183.

[29] Ward, W. J.; Ritzer, A.; Carroll, K. M.; Flock, J. W. *J. Catal.* 1986; vol. 100, pp 240-249.

[30] Gasper-Galvin, L. D.; Sevenich, D. M.; Friedrich, H. B.; Rethwisch, D. G. *J. Catal.* 1991; vol. 128, pp 468-478.

[31] Colin, P.; Garcia-Escomel, C.; Masteau, J. US Patent 8,030,510B2, October 4, 2011.

[32] Lewis, L. N.; Ward, W. J. *Ind. Eng. Chem. Res.* 2002, vol. 41, pp 397-402.

[33] Ehrich, H.; Born, D.; Richter-Mendau, J.; Lieske, H. *Appl. Organometal. Chem.* 1998; Vol. 12, pp 257-264.

[34] Voorhoeve, R. J. H.; Geertsema, J. H.; Vlugter, J. C. *J. Catal.* 1965, vol. 4, pp 43-55.

[35] Voorhoeve, R. J. H.; Lips, J. A.; Vlugter, J. C. *J. Catal.* 1964, vol. 3, pp 414-25.

[36] Nakayama, H.; Nakanishi, T.; Tsukioka, K.; Satoh, Y., EP Patent 0869130A2, April 1998.

[37] Reed, C. E.; Coe, J. T., US Patent 2389931A, Nov. 1945.

[38] Williamson, O. R.; Strange, W. T.; Fritze, S. R.; Lanthier, C. P., EP Patent 0256876A2, Aug. 1987.

[39] Baker, Thomas H. Jr. U.S. Patent 4585643, April 29, 1986.

[40] Padovani, F. A.; Miller, M. B.; Moore, J. A.; Fowler, J. H.; June, M. N.; Matthews, J. D.; Morton, T. R.; Stotko, N. A.; Palmer, L. B. U.S. Patent 4092446, May 30, 1978.

[41] Jiao, Y.; Salce, A.; Ben, W.; F.; Ji, X.; E.; Lynch, D. JOM, 2011, pp 28-31.

[42] Ceccaroli, B.; Lohne, O. in Handbook of Photovoltaic Science and Engineering, eds. Loque A.; S.; Wiley, West Sussex, UK, 2003, pp 153 – 204.

[43] Schmid Silicon Technology (Sept. 20, 2011), The Advantages of the Schmid Process [www page]. http://www.schmid-silicon.com/en/technology/schmid-process-vs-siemens-process.html .

[44] Ingle, W. M., Peffley, M. S., Nagaraja, H. S., U.S. Patent 4526769, July 2, 1985.

[45] REC (March 31, 2011). Annual Report 2010 [www page]. http://www.rec group.com/view?feed=R/136555/PR/201103/1502158.xml

[46] Gautreaux, M. F.; Allen, R. H. US Patent 4,784,840, Nov. 15, 1988.

[47] Mulla, I.; Choube, A.; Dongare, M.; Sinha, A. *Indian J. Chem. Sect* A., 1989,vol. 27 pp 756-758.

[48] D. Petrov, B. F. Mironov, V. A. Ponomorenko, and R. A. Cherpyshev, *Synthesis of Organosilicon Monomers*, ConsultantsBureau, New York, 1964.
[49] D. T. Hurd, *J. Am. Chem. Soc.*, 1945, vol. 67, pp 1545-1548.
[50] Hatcher, D. B. U.S. Pat. 2,458,703, Jan. 11, 1949.
[51] Bailey, D. L. U.S. Pat. 2,745,860 , May 15, 1956.
[52] Yamada, H.; Ogawa, H.; Hosokawa, T.; Tachikawa, M. JP 63/170210 A2 [88/170210] , July 1988.
[53] Consultant on Demand: Engineering Assistance to Industry (accessed 04/14/12). Boron and Phosphorus Removal/Impurity Reduction [www page]. http://consultant-on-demand.net/BoronPhosphorus.htm
[54] Goldstein, M. in Metallization: Interconnects, University of Arizona, Tucson, Arizona, Spring 2011.
[55] Philipossian, A. in Chemical Vapor Deposition, University of Arizona, Tucson, Arizona, Spring 2011.
[56] Time Domain CVD, silane/Oxygen Thermal CVD [www page]. http://www.time domaincvd.com/CVD_Fundamentals/films/SiH4_O2_thermal.html (accessed April 2, 2012).
[57] VanOmmeren, J. in *The American Institute of Chemical Engineers*, Allentown, 2011.
[58] Roigelstad, T.; Mosovsky, J.; Valdes, J.; Lichtenwalner, C. P. (1994)
[59] Silane Safety Improvement S71 Final Report [pdf]. http://www.sematech.org/docubase/document/2405aeng.pdf
[60] Hirano, T. J. of Loss Prevention in the Pprocess Industries, 2004, vol. 17, pp 29-34.
[61] Fthenakis,V. M. (2008), Summary Review of Silane Ignition Studies [www page]. http://www.bnl.gov/pv/abs/abs_149.asp .
[62] D. Biello (April 2, 2010), Explosive Silicon Gas Casts Shadow on Solar Power Industry [www page]. http://www.scientificamerican.com/article.cfm?id=explosive-gas-silane-used-to-make-photovoltaics
[63] Mather, K (March 22, 2012), At least two people injured in chemical explosion at Northwest Portland factory [www page]. http://www.oregonlive.com/portland/index.ssf/2012/03/at_least_one_person_injured_in.html .
[64] Ngai, E. Y. (2009), Silane Safety [pdf] http://seshaonline.org/regions/ssanorthtexasc/2009_SilaneSafety.pdf .
[65] Ngai, E. Y.; Fthenakis, V. (20080), Silane Safety/Lessons Learned [pdf]. http://www.docstoc.com/docs/47529562/Silane-SafetyLessons-Learned-and-Accident-Prevention#
[66] 3E Company, MSDS Free Search [WWW page]. www.msds.com (accessed 03/18/12).
[67] Environmental Protection Agency. (2007). Acute Exposure Guideline Levels for Silane [PDF file]. http://www.epa.gov/oppt/aegl/pubs/silane_interim_nov_2007.v1.pdf
[68] Wikipedia, Median Lethal Dose [WWW page]. http://en.wikipedia.org/wiki/Median_lethal_dose.
[69] Perry, R. H.; Green, D. W. Perry's Chemical Engineerings *Handbook*, 8th ed.; McGraw-Hill: New York, NY, 2008.
[70] PCR Inc.; Research Chemicals Catalog 1990-1991, PCR Inc., Gainesville, FL, 1990; 1.
[71] Tannenbaum, S.; Kaye, S.; Lewenz, G. F. J Am Chem Soc. 1953, vol. 75, 3753-3757.
[72] ChemSpider. http [WWW page]:http//www.chemspider.com

[73] Chemical Book [WWW page]. http://www.chemicalbook.com/CASDetailList_0_EN.htm

[74] Drefahl, A. Quantitative Property-Property Relationships (QPPRs) and Molecular-Similarity Methods for Estimating Flash Points of Si-Organic and Ge-Organic Compounds, Proceedings of the iEMSs 3rd Biennial Meeting, Burlington, USA, July 2006.

[75] Stull, D. R. *Ind. Eng. Chem.* 1947, vol. 39, pp 517-540.

[76] Hshieh, F.Y. *Fire Mater.* 1997, vol. 21, pp 277-282.

[77] Stanford Linear Accelerator Center. Physical Constants of Inorganic Compounds. [PDF file]. http://www.slac.stanford.edu/BFROOT/www/Detector/Backgrounds/BkG4Sim/Planning/Validations/neutronCounters/04_02_85.pdf

[78] Doncaster, M.; Walsh, R. T. *J Chem Soc Faraday* 1986, 82, pp 707-717.

[79] Chase, M. W. NIST-JANAF Thermochemical Tables, 4th ed.; American Institute of Physics for the National Institute of Standards and Technology: New York, NY, 1998.

[80] Steele, W. V. *J. Chem. Thermodyn* 1983, 15, pp 595-601.

[81] Voronkov, M. G.; Klyuchnikov, V. A.; Danilova, T. F.; Danilova, T. F.; Pepekin, V. I.; Korchagina, A. N.; Khudobin, Yu. I. *Russ. Chem. B* 1986, vol. 35, pp 1790-1795.

[82] Rappoprt, Z.; Apeloig, Y. *The Chemistry of Organic Silicon Compounds*, Wiley: West Sussex, 2006; Vol. 2.

[83] NIST Chemistry WebBook. [WWW page]. http://webbook.nist.gov/chemistry/

[84] Voronkov, M. G.; Baryshok, V. P.; Klyuchnikov, V. A.; Danilova, T. F.; Pepekin, V. I.; Korchagina, A. N.; Khudobin, Yu. I. *J Organomet Chem.* 1988, 345, pp 27-38.

[85] Saul Patal, Zvi Rappoport. *The Chemistry of Organic Silicon Compounds: Part I.*; John Wiley: New York, NY, 1989.

[86] Moedritzer, K. *Adv Organomet Chem.* 1968, 6, pp 171-263.

[87] Hajiev, S. N.; Agarunov, M. J. *J. Organometal Chem.* 1970, 22, pp 305-311.

[88] Schlegel, H. B. *J. Phys. Chem.* 1984, vol. 88, pp 6254-6258.

[89] Walsh, R. Thermochemistry of Silicon-Containing Compounds. Part 1: Silicon–Halogen Compounds, An Evaluation. *J. Chem. Soc. Faraday* 1983, vol. 79, pp 2233-2248.

[90] Ho, P.; Coltrin, M. E.; Binkley, J. S.; Melius, C. F. *J. Phys. Chem.* 1985, vol. 89, pp 4647-4654.

[91] Ho, P.; Melius, C. F. *J. Phys. Chem.* 1990, vol. 94, pp 5120-5127.

[92] Keister, J. W.; Tomperi, P.; Baer, T. J. Am. Soc. Mass Spectrom. 1998, vol. 9, pp 597-605.

[93] Potzinger, P.; Lampe, F. W. *J. Phys. Chem.* 1970, vol. 74, pp 719-724.

[94] Wong, H-W.; Nieto, J. C. A.; Swithart, M. T.; Broadbelt, L. J. *J. Phys. Chem.* A. 2004, vol. 108, pp 874-897.

[95] Katzer, G.; Ernst, M. C.; Sax, A. F.; Kalcher, J. *J. Phys. Chem.* A. 1997, vol. 101, pp 3942-3958.

[96] Hidding; Fikri, M.; Bozkurt, M.; et al. *Ttans. JPN Soc Aeronaut S.* 2010, vol. 8, pp 39-45.

[97] Roine, A. HSC Chemistry 6.0, Outokumpu, Pori, Finland, 2006.

[98] Richardson, F. D. Physical Chemistry of Melts in Metallurgy Vol. 2; Academic Press: London, UK, 1974;pp 328-346.

[99] Sullivan, T. Finding the Solar Market's Nadir, LUX Research Report LRSI-R-09-001, Feb. 2009.

[100] Hoffmann, W.; Pietruszko, S. M.; Viaud, M. (2004). Towards an Effective European Industrial Policy for PV Solar Electricity [pdf]. http://s3.amazonaws.com/zanran_storage/masstech.org/ContentPages/29583216.pdf .

In: Silane: Chemistry, Applications and Performance
Editors: Katsuo Moriguchi and Susumu Utagawa
ISBN: 978-1-62257-432-2

Chapter 3

SILANES FOR ADHESION PROMOTION AND SURFACE MODIFICATION

***Christie Ying Kei Lung*[*] *and Jukka Pekka Matinlinna*[≠]**
Dental Materials Science, Faculty of Dentistry, The University of Hong Kong, Hong Kong SAR, P.R. China.

ABSTRACT

Silanes are synthetic compounds containing one or more Si-C bonds. They have wide application in industry, dentistry and medicine. They are used mainly as surface coating agents for surface modification and as coupling agents for adhesion promotion between two dissimilar materials. The role of silanes as either surface coating or coupling agents depends on their chemical structure. A wide variety of reactive organic functional groups can be bonded to the silicon atom such that silanes can react with the functional groups of many substrates. When silanes are used as a surface coating agent, the physical and chemical properties of the substrate surfaces are improved. As coupling agents, silanes enhance the chemical bonding between two dissimilar materials. Reaction mechanisms to illustrate these actions of silanes will be presented. Some techniques of surface treatment such as air-abrasion and chemical etching to enhance the adhesion between a silane and the substrate surfaces will also be covered. Findings from current studies of various silanes as surface-coating and coupling agents in different areas, with an emphasis on dentistry, will be discussed. The focus will be on the findings of performance of silanes as surface coating and coupling agents in different aspects of applications, including limitations in use. Finally, an outlook about on future developing trends of silanes in dental materials science will be provided. (219 words)

[*] Corresponding author: Christie Ying Kei Lung, Dental Materials Science, Faculty of Dentistry, The University of Hong Kong, 4/F, Prince Philip Dental Hospital, 34 Hospital Road, Sai Ying Pun, Hong Kong SAR, P.R.China, Tel.: +852 2859 0209; Fax 852 2548 9464, E-mail address: yklung@graduate.hku.hk.

[≠] E-mail address: jpmat@hku.hk (Jukka Pekka Matinlinna).

1. INTRODUCTION

Silicon is a group XIV element. Group XIV members include carbon, silicon, germanium, tin, and lead. Among these members, carbon and silicon are the two most important elements in living organisms and are used in many applications. Silicon makes up a large variety of chemical compounds such as silica (silicon dioxide, SiO_2), silicates, silicones, silicon carbide and silane, *etc.* All of these silicon compounds have wide application in many areas. Silica is used in manufacturing of glasses and ceramic bricks. Silicates are used in cement and concrete in construction materials. Silicon carbide, SiC, is used in abrasive and cutting tools, automobile parts and heating elements. Silicone is used in sealants, adhesives, lubricants, silicone liners in technology [1].

Silanes are one of the most important classes of silicon compounds and contain one or more direct Si-C/Si-H bonds. Silanes and their derivatives are widely used in many applications. Within the silane family, organosilanes are the most important members.

Organosilanes which are not found in nature are mainly synthesized from silica which is very abundant on Earth. They are used primarily for surface modification and adhesion promotion. Surface modification is to alter the physical, chemical and biological properties of the substrate surface. Numerous applications require the connection of dissimilar inorganic and organic materials to fulfil certain purposes. However, the interaction is very weak at the interfacial layer that a coupling agent is required to connect them. Silanes that have two or more different functional groups, (one functional group attached to silicon atom) can connect dissimilar materials together by forming chemical bonds.

There are several methods for applying silanes onto different substrates. Functional silanes can be prepared by dissolving the silane in a solvent mixture of ethanol and water at pH 4.0-5.0. The substrate is immersed in the solution and then taken out for drying. A thin layer of silane film can also be applied on a substrate surface by brushing. It may also by spraying the silane solution onto the substrate surface and allowed to dry and curing. Another method of silane coating is by chemical vapour depositions. In a closed chamber, the silane reservoir is heated under reduced pressure, and silane vapour becomes deposited onto the substrate surface. Silanes can also be applied by spin coating. Silane solution is deposited by spinning the substrate at low speed, and then the substrate is rinsed [2].

In this chapter, the aspects of silane chemistry, the silane application in dentistry, medicine and industry and the future development trends of silanes are discussed in details.

2. SILANE CHEMISTRY

2.1. Structure and Reactivity

The action of silanes as coating or coupling agents depends on the chemical structure. The general formula for functional silanes is: $Y–(CH_2)_n–SiR_3$. The hydrolysis of $-SiR_3$ will form –SiOH, which can bond to the surface hydroxyl groups of inorganic substrates. By varying the nature of functional group Y, silanes can act as surface-modifying agents and coupling agents. For example, if Y is a long hydrocarbon chain, the coating of the silane on the substrate surface would enhance the hydrophobicity of the substrate surface. If Y is NH_2-

or >C=C<, which would react with epoxy resins or vinyl and acrylate resins, these functional silanes would act as a coupling agent to connect resin to an inorganic substrate surface [1, 2].

The effectiveness of adhesion of silane coupling agent varies with different substrates (Fig. 1). A strong siloxane bond (-Si-O-Si-) is formed between silane coupling agents and silica, quartz, or glass through the condensation of silanol groups on surface hydroxyl groups. For metals and alloys, a thin layer of oxide is formed on the surface which is also hydrated at ambient temperatures. The Si-O-M bond is formed after condensation of silanol groups with the hydroxyl groups of the hydrated surfaces. The bond strength varies with different metals and alloys. Several factors affect the bond stability: 1) the isoelectric point (IEP) of the surface. IEP is defined as the pH where the zeta potential of the surface is zero, 2) the ionic character of the metal oxide, and 3) the solubility of the metal oxides in water [1]. For substrates such as gypsum, graphite and carbon black, which all contain very low concentrations of surface hydroxyl groups, the adhesion of silane coupling agents is therefore poor [3, 4].

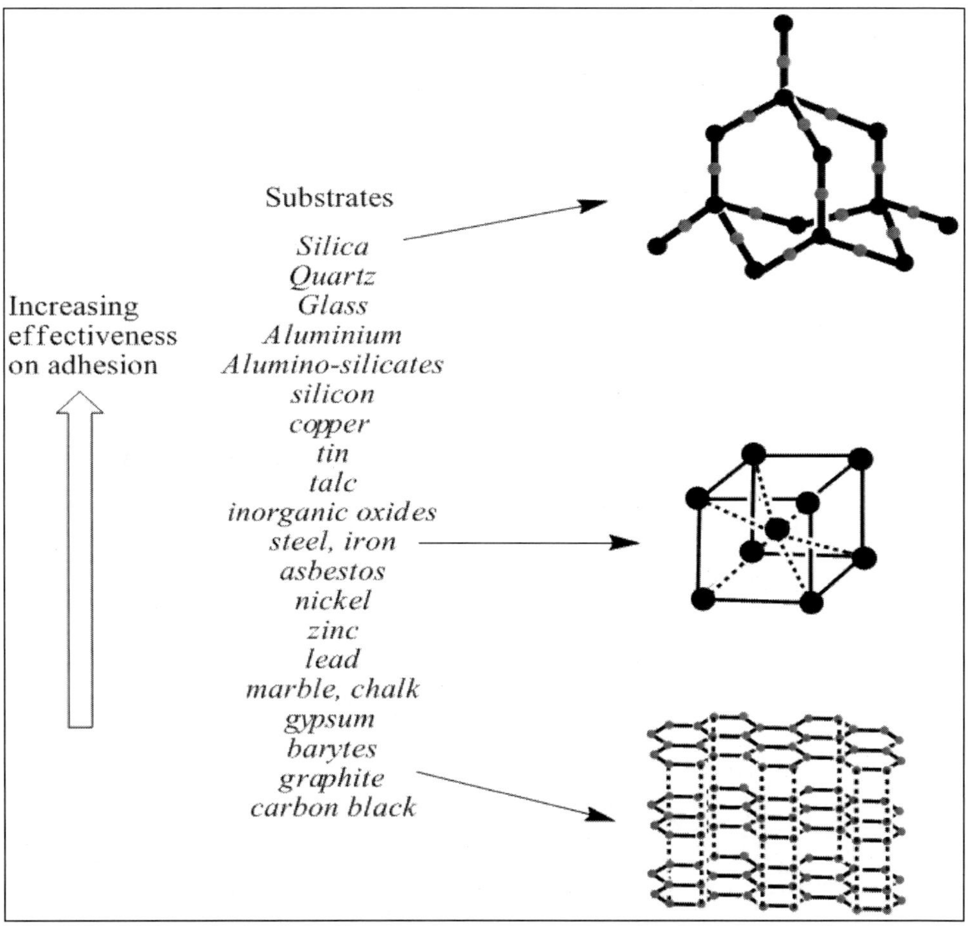

Figure 1. Effectiveness of silane adhesion to different substrates [modified from ref. 2].

2.2. Silicon Compounds and Carbon Compounds

Silicon has several similarities to, but also several differences from the first member of group XIV. The electronic configuration of carbon is $1s^2 2s^2 2p^2$ and that of silicon is $1s^2 2s^2 2p^6 3s^2 3p^2$. Both atoms have four outermost electrons which form four-coordinate tetrahedral compounds with sp^3 hybridization. Hence, most of the compounds of carbon and silicon are four-coordinated. However, carbon is a second-period element and silicon is a third-period element. Because of the presence of the low-lying vacant 3d orbital in third-period elements, silicon can expand the "octet rule" to form five- and six-coordinate compounds with sp^3d and sp^3d^2 hybridization with most electronegative ligands [5, 6]. Silicon is also less electronegative than carbon. Thus, the bond polarity of silicon compounds is higher than that in analogous carbon compounds. This increased bond polarity also accounts for the difference in chemical reactivity of silicon and carbon compounds. For example, the rates of hydrolysis for most organo-carbon compounds are slower than for organo-silicon compounds without any acid or base catalysts. Owing to the larger atomic size of Si, the formation of the Si=Si bond is less favorable than that of the C=C bond (because of poorer π overlap of 3p orbitals than 2p orbitals) [7, 8]. Formation of the Si=Si bond is favorable when bulky substituent groups such as mesityl (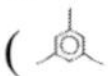) are attached to prevent polymerization of monomeric silicon compounds [8]. Carbon can form long chain compounds with other carbon atoms. The ability for silicon to form long chain compounds is lower than carbon. This may be attributed to the larger atomic size of silicon such that the sigma bond overlap of 3p orbital is less effective than that of 2p orbital of carbon.

2.3. Synthesis of Silanes

Organosilanes do not occur naturally. They are produced from silica, a very abundant mineral in earth [9]. Silica is reduced to silicon through a reaction with carbon under very high temperature according to the following equation (1):

$$SiO_2(s) + C(s) \xrightarrow{\Delta} Si(s) + CO_2(g) \tag{1}$$

Then, trichlorosilane is synthesised by the direct reaction of silicon with hydrogen chloride in the presence of a copper catalyst at 300°C [10]:

$$Si(s) + 3HCl(g) \xrightarrow{\Delta} HSiCl_3(g) + H_2(g) \tag{2}$$

The next step is the synthesis of 3-chloropropyltrichlorosilanes and the final step produces 3-chloropropyltrialkoxysilanes synthesis [11]:

$$ClH_2CHC{=}CH_2(g) + HSiCl_3(l) \longrightarrow ClH_2CH_2CH_2SiCl_3(l) \tag{3}$$

$$ClH_2CH_2CH_2SiCl_3(l) + 3\ ROH(l) \longrightarrow ClH_2CH_2CH_2Si(OR)_3(l) + 3\ HCl(g) \qquad (4)$$

Because the -Cl can be converted into other functional groups [11], the alkoxysilane can act as either a coupling agent or a surface-modification agent.

Organochlorosilanes can also be prepared by the reaction of tetrachlorosilane with organozinc [12]:

$$SiCl_4 + n/2\ ZnR_2 \longrightarrow R_nSiCl_{4-n} + n/2\ ZnCl_2 \qquad (5)$$

Organosilanes are non-toxic to humans and are environmental friendly. However, the hydrolysis of methoxysilane produces methanol which is toxic. In some cases, methoxysilane is replaced by ethoxysilane.

2.4. Reaction Mechanisms of Silanes

2.4.1. Silane Activation

Organofunctional silanes contain two different reactive functional groups that can react and couple with various inorganic and organic materials. Thus, they act as adhesion promoters to join dissimilar materials. These functional silane coupling agents are first activated by hydrolysis ($\equiv$SiOR $\rightarrow$ $\equiv$SiOH) before they can bond, via hydroxyl groups on the substrate surface. The first step of hydrolysis of silane to silanol ($\equiv$SiOH) takes place by the fast and reversible protonation of the alkoxy group of the silane at pH 4. Then, it undergoes a bimolecular nucleophilic substitution (S_N2) reaction at the central silicon atom. This first involves the backside attack by a water molecule (nucleophile) to the central silicon atom (electrophile), to give a penta-coordinate trigonal bipyramidal transition state. A new bond is then formed between the nucleophile and the silicon centre, and a bond is cleaved between the leaving group (alcohol) and the silicon centre. Finally, this gives the product with inversion of configuration after the hydrolysis process [13]. A suggested reaction mechanism for silane hydrolysis is shown in Fig. 2. Two factors affect the rate of silane hydrolysis: steric and inductive effects of alkoxy groups on the silane, with the steric effect as the dominant factor [14]. As depicted in Fig. 2, the nucleophile attacks from the behind at the opposite side of the leaving group. As the alkoxy group is more bulky, steric repulsion is increased when the nucleophile approaches the central silicon atom. Accordingly, the activation energy of the transition state is increased which results in a reduction of the rate of hydrolysis. This effect may be illustrated by using space-filling model of silane molecules (Fig. 3). The approach of the nucleophile, water, to the central silicon atom is more difficult in triproxyvinylsilane than in trimethoxyvinylsilane as the silicon centre is buried more deeply in the former silane molecule.

2.4.2. Silane Surface Modification and Adhesion

Hydrophobic silane coupling agents produce hydrophilic species containing hydroxyl groups after hydrolysis. When the silane solution is applied onto the substrate surface at ambient temperature, the free silanol groups first form hydrogen bonds with the hydroxyl (-

OH) groups on the inorganic substrate surface. Then, a –Si–O–(substrate)– linkage is formed between the silanol and "HO-(substrate)" surface after condensation. Surface treatment with mineral acids such as hydrochloric acid and nitric acid could enhance the formation of surface hydroxyl groups which in turn increases the degree of bonding between silane coupling agents and the hydroxylated substrate surface [15].

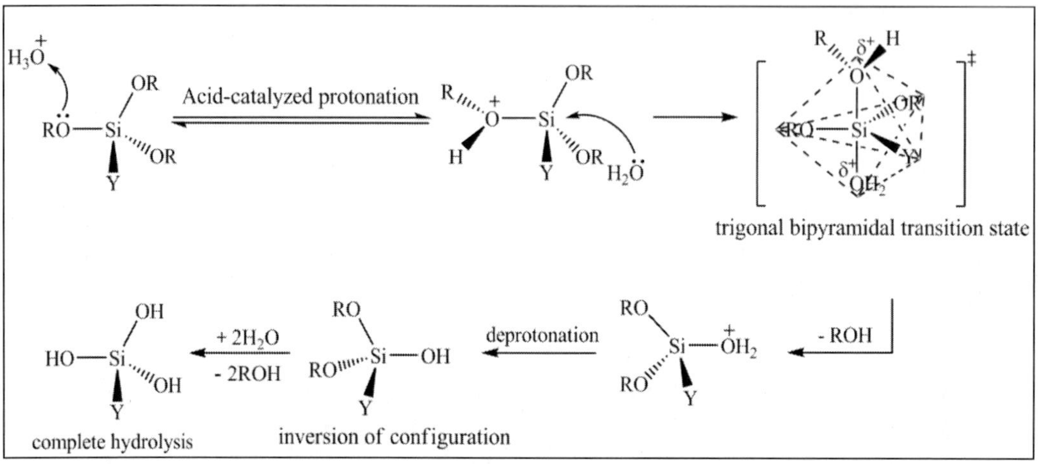

Figure 2. Reaction mechanism of silane hydrolysis.

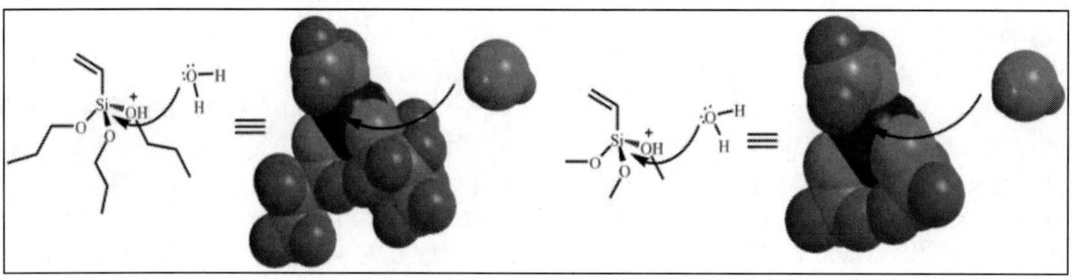

Figure 3. The steric effect of alkoxy groups on silane hydrolysis using space-filling models of tripropoxyvinylsilane and trimethoxyvinylsilane for illustration.

The non-functional hydrocarbon chain of silane coupling agents would increase surface hydrophobicity by changing surface energy of the substrate and thus the substrate surface is modified. Resin composite consists of functional groups, *e.g.* >C=C< double bonds that can react with the organofunctional groups, *e.g.* >C=C<, of silane coupling agents.

The initiator components in the resin composite decompose into high-energy and reactive free radicals by heat or light activation.

These reactive species react with the >C=C< bond in both resin monomer and silane molecules to form new free radical species. Addition of these species between resin composite monomers and silane molecules forms new C-C bonds. A linkage is thus formed between resin composite and the inorganic substrate surface through the silane coupling agent (Fig. 4) [14].

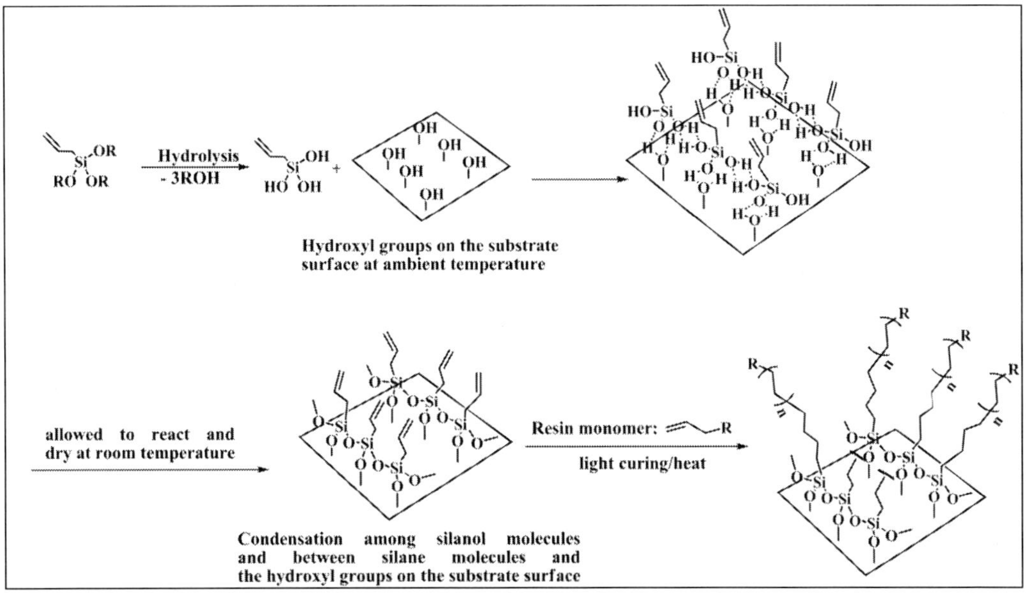

Figure 4. Action of silane in adhesion promotion between resin and inorganic substrate surface.

3. Surface Conditioning of Some Dental Materials

Surface conditioning of dental materials is a surface treatment that increases the surface roughness, *i.e.* the surface energy and also creates micropores for infiltration of silane coupling agents and resin composites [16]. An increased surface energy results in better wetting for bonding [17]. Some of the common surface-conditioning methods for zirconia [18], metals (such as titanium [19]), base and noble metal alloys and porcelain (including repair of fractured porcelain [20]), as all applied in dental clinical practice are discussed below.

3.1. Air Abrasion Techniques

In dental laboratories, the air abrasion process is carried out by blasting surfaces with alumina particles of an average size of 50 μm under compressed air at a pressure of 380 kPa, for around 10-15 s at a perpendicular short distance (such as 10 mm) from the nozzle to the surface [21]. The impact of high-energy particles onto the substrate surface increases the surface roughness which in turn enhances the mechanical interlocking for bonding [22].

3.2. Tribochemical Silica Coating

A tribochemical Rocatec™ System (3M ESPE, Seefeld, Germany), using silica-coated alumina particles, Rocatec™ Soft (30 μm silica-coated alumina) or Rocatec™ Plus (110 μm silica-coated alumina) is used for the surface conditioning of various substrates such as ceramics, metals and metal alloys. As with air abrasion, the substrate surface is blasted at a pressure of 280 kPa under compressed air at a perpendicular distance of around 10 mm for 30 s/cm^2 [23]. The transfer of kinetic energy from the impact of the sand particles on the substrate surface causes the melting of the surface microscopically [24]. The silica-coated alumina particles penetrate and bond to the substrate surface, so that a thin silica layer is coated onto it. The silica-coated surface is further treated by coating it with a thin layer of silane coupling agent. Then, it is cemented using resin composite cement to complete the dental restoration. Nowadays, this is the most popularly used method in dental practice for the surface conditioning of dental restorative materials.

3.3. Chemical Treatments

Porcelains are used in ceramic restorations owing to their superior esthetic appearance and metal-free substructure. Dental porcelain is first etched with hydrofluoric acid, HF, followed by silanization and finally cementation as a dental restoration [20]. When porcelain is treated with HF, tetrafluorosilane, SiF_4, is produced which then reacts with HF to form a soluble hydrofluosilicic acid, H_2SiF_6. The glassy matrix is dissolved and can be washed away, leaving a microscopically porous and high energy surface [25]. This can enhance mechanical retention for bonding [26].

Another chemical surface treatment method is to coat the substrate surface with a silica layer by using a PyrosilPen™ device (SurA Instruments, Jena, Germany). A silane, tetraethoxysilane, decomposes in the flame and forms SiO_x-C fragments which become deposited onto the surface, giving a thin “glasslike” layer. The surface is then treated with silane coupling agent, followed by cementation [27].

4. Silane Application

4.1. Dentistry

In dental restorations, tribochemical silica-coating using a Rocatec™ system followed by silanization is a pre-treatment step before cementation. A commonly used silane is the commercial silane primer 3-methacryloxyproyltrimethoxysilane [28]. It is pre-hydrolysed in ethanol-water solvent mixture before application. However, pre-hydrolysed silane solution has relatively short shelf life. It may turn milky after use and cannot be used. There are also two-bottle silane systems on the market. These systems consist of an unhydrolyzed silane in ethanol in one bottle and an aqueous acetic acid solution in the other bottle [29]. The two solutions are mixed to allow silane hydrolysis just before use. This procedure can increase the shelf life of silane coupling agent.

In dental research, the adhesion of a resin composite to restorative materials is evaluated by shear bond strength testing and micro-tensile bond strength testing. These methods are based on application of a load at the adhesive interfacial layer between resin composite and restorative materials until facture occurs (Fig. 5). The higher the recorded bond strength values are, the stronger the adhesion is at the interfacial layer.

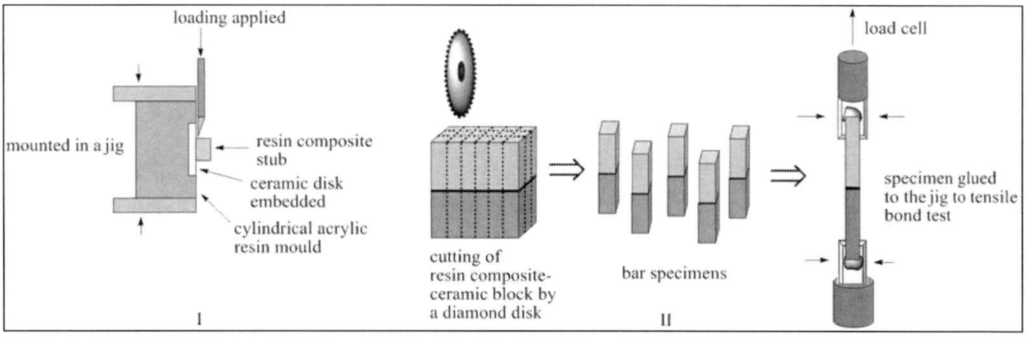

Figure 5. Schematic diagram of (I) Shear bond strength testing and (II) Micro-tensile bond strength testing.

Silanes perform well in adhesion promotion for silica-based restorative materials such as porcelain [20]. However, for non-silica based materials such as titanium [30], metal alloys [31] and zirconia [32], surface pre-treatment of the substrate surfaces by silica coating is required so that silanes can form durable siloxane linkages with these materials. The application of silanes in dentistry will be discussed in the following sections.

4.1.1. Ceramics

Ceramic materials such as zirconia, porcelain and glass-infiltrated alumina are used as crowns, onlays, inlays, and bridges in dental restorations and as brackets in orthodontics. The surface of the materials is pre-treated by sandblasting for non-silica based materials and acid etching for silica-based materials, followed by silanization and finally cementation. Dental ceramics such as used for crowns and bridges may fracture owing to intraoral occlusal forces, design and manufacturing faults and impact and fatigue load. For such fractures, repair is more economical and time-saving than manufacture of new restorations. The repair involves four steps in general: 1. surface roughening, 2. sandblasting, 3. acid-etching and 4. silanization followed by cementation [33].

4.1.2. Metals and Base Metal Alloys

Titanium is used in dental subgingival implants [34] and chromium-cobalt alloys are used in metal to ceramic restorations [35]. Surface pre-treatment by tribochemical silica coating for these metals/metal alloys before silanization and cementation is necessary to allow durable bonding, as these are all non-silica based materials. For noble metals and metal alloys, besides using silane coupling agents for adhesion promotion, special primers can be used as alternatives to silane coupling agents for adhesion promotion. Examples of these primers are phosphate for base metal alloys and thiol for noble metals and noble metal alloys [36].

4.1.3. Resin Composite Fillers

Fillers such as silica, hydroxyapatite, boron silicates and alumina are added in dental resin composites to improve their physical and mechanical properties of resin [37]. Moreover, barium and strontium glass fillers are added to resin composite to give them X-ray opacity to facilitate radiological monitoring *in vivo* [38]. Incorporation of silanized fillers could further enhance the mechanical properties of the resin composites than non-silanized fillers [39].

Silane coupling agents can form linkage between the fillers and the resin which in turn increase the bonding.

4.1.4. Fiber-Reinforced Resin Composites

Applications of fiber-reinforced composites are found in endodontics, periodontology, prosthodontics and cariology, [40]. Silanized glass fibers are added in acrylic resin to reinforce the denture base by strengthening the interaction between the fiber and resin composite. The E-glass fibers are pre-treated with silane coupling agents before being incorporated into the resin matrix. Fiber-reinforcements are also found in denture base materials that are used in fixed partial dentures and removable prosthodontics.

4.1.5. Current Studies

In dental practice, the surface treatment for non-silica-based restorative materials is usually done by tribochemical silica coating. Other methods of silica coating are under investigation. Current studies in this area include laser irradiation, selective infiltration etching, and chemical vapour deposition of silianes onto the substrate to enhance the bonding [41-43]. All these methods have different mechanisms to activate the zirconia surface for adhesion promotion to resin composite.

Table 1. Some functional silanes studied in resin zirconia/titanium bonding system

Silane studied	Bonding test system	Refs
3-acryloxypropyltrimethoxysilane	Resin-zirconia/titanium	44, 45, 46, 47, 51, 53, 54
3-glycidoxypropyltrimethoxysilane	Resin-zirconia/titanium	44, 47
3-isocyanatopropyltriethoxysilane	Resin-zirconia/titanium	23, 45, 47, 48
3-isocyanatopropyltrimethoxysilane	Resin-zirconia	23
3-methacryloxypropyltrimethoxysilane	Resin-zirconia/titanium	44, 45, 47, 48, 49, 50, 52, 54
3-mercaptopropyltrimethoxysilane	Resin-zirconia/titanium	49, 51
strylethyltrimethoxysilane	Resin-titanium	47
tris(3-trimethoxysilylpropyl)isocyanurate	Resin-titanium	50
[3-(triethoxysilyl)propyl]urea	Resin-zirconia/titanium	52
bis-[3-(triethoxysilyl)propyl]polysulfide	Resin-titanium	51
bis-[3-(triethoxysilyl)propyl]tetrasulfide	Resin-titanium	44
N-[3-(trimethoxysilyl)propylethylenediamine]	Resin-titanium	51, 52
tetrakis-(2-methacryloxyethoxy)silane	Resin-titanium	44
vinyltriisopropoxysilane	Resin-titanium	50
functional silanes + *bis*-1,2-(triethoxysilyl)ethane	Resin-zirconia/titanium	44, 49, 52, 53, 54
3-acryloxypropyltrimethoxysilane + *bis*[3-(trimethoxysilyl)propyl]amine	Resin-zirconia	53

Studies investigating the effects of different silane coupling agents on bonding of resin-composite to restorative materials report bond strength degradation over time under artificial aging. This effect may be due to the hydrolytic cleavage of siloxane bonds at the interfacial layer, which lowers the bond strength [44]. Extensive studies have been done or are still undergoing to investigate the effect of different silanes on the bond strength and hydrolytic stability of the bonding between resin composite and restorative materials bonding. These silane coupling agents contain different functional groups. Some functional silanes that have been studied are listed in Table 1. Bond strength values for dry storage, and for under-water storage and thermocycling vary by different extents for silanes with different functional groups. These findings can be explained by the fact that the reactivity towards resin composites varies according to the different functional groups present in the silanes. Because the addition of cross-linking silanes to functional silanes results in an extensive siloxane network that connects the functional silanes, the use of cross-linking silanes could minimize the hydrolytic cleavage of siloxane linkage at the interfacial layer [54].

4.2. Medicine

4.2.1. Medical Implants

When people have severe hip damage, the hip may be replaced by an artificial one that is composed of two parts: the femoral head and neck and the acetabulum cup. The femoral head and neck is made of metal alloy (Co-Cr or $TiAl_6V_4$), and the acetabulum cup is made up of polyethylene [55].

The two parts are held in place by poly(methylmethacrylate) (PMMA) bone cement. However, the main problem of hip replacement (arthroplasty) is the aseptic loosening between the metal stem surface and PMMA bone cement. This adhesion failure is due to the moisture rather than the body fluid which causes the debonding at the metal stem-PMMA interface. The water molecules diffuse through the bone cement to the bonding interfacial layer and cause progressive bond breaking even without mechanical loading [56].

There are two approaches to increase the bond hydrolytic stability. The first approach is to pre-coat the metal stem surface with a low viscosity PMMA such that the surface roughness of the metal stem is increased, which improves the mechanical bonding [57]. Muller *et al.* [57] and Raab S *et al.* [58] reported that the hydrolytic stability of the bonding metal to PMMA bone cement by pre-coating with thin film PMMA was significantly higher with a pre-coating of a thin film of PMMA than without a pre-coating.

The bond strengths for specimens without pre-coating were significantly lower after immersion in saline solution than after dry storage. The bond strengths of specimens pre-coated with PMMA were not decreased significantly after immersion in saline solution.

The second approach is to coat the metal stem surface with a silane coupling agent [59]. The bonding between the silane coupling agent and the metal stem is weak. However, the surface can be modified first, such that strong bonding is formed between the silane coupling agent and the surface of the metal stem. This can be achieved by coating a thin layer of silica onto the surface. The silane can form strong siloxane bond (-O-Si-O-) bonds with the silica layer. The silica layer is deposited onto the metal stem surface by chemical vapor deposition method or by sandblasting.

A suggested mechanism by which the silane coupling agent then bonds the silica-coated metal stem to PMMA bone cement is shown in Fig. 6. Studies [56, 60] have shown that the bond degradation after immersion in saline solution at 37°C for different storage times was significantly reduced.

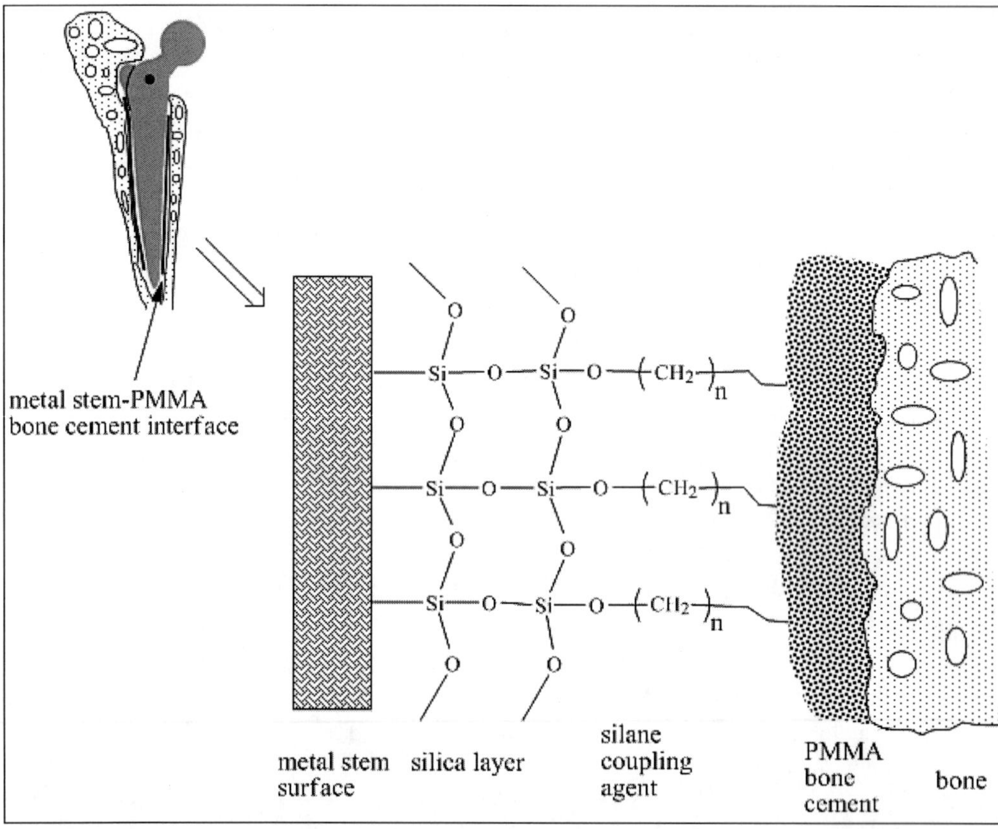

Figure 6. Mechanism by which silane coupling agents connect the metal stem of a replacement hip to PMMA bone cement.

4.2.2. Biological Immobilization

Biological immobilization is a widely used technology to transform the "inert" substrate into a biologically active site for specific molecular recognition. Functional agents are used to modify the inorganic or organic inert solid supports such as ceramics, minerals, polyamides, and polysaccharides, to allow them to bind biological agents so that they have various biological properties [61]. Several applications of silanes in this area are discussed below.

4.2.2.1. Antimicrobial Activity

Many materials used in medical devices such as biomedical implants and transcutaneous devices are exposed to various microorganisms that bind to the surface and thus affecting the device's functionality. When the number of microbial cells is high, biofilms are formed on the surface of the medical devices [62, 63]. These biofilms can lead to serious infection as they have high antibiotic resistance. In order to prevent bacterial growth and biofilm formation, the surfaces of devices can be modified with functional silanes that can bind to functional agents that have antimicrobial activity. Most amphiphilic peptides have antimicrobial activity. They can also couple to the functional silane-modified substrate by binding to the functional group of silane such as epoxy. The immobilization of amphiphilic peptides has good selectivity to deplete bacterial cells [61, 64].

4.2.2.2. Drug Delivery

Superparamagnetic nanoparticles can be used as drug carriers for controlled drug release, for example, to kill tumor cells [65]. However, after their injection, these nano-particles agglomerate and absorb plasma proteins because of their large surface area to volume ratio. They are then cleared by macrophages before they reach the target cells [66]. One approach to solve this problem is to modify the surfaces of nanoparticles with poly(ethylene glycol) (PEG). When coated with PEG, the nanoparticles are non-immunogenic, protein resistant and non-antigenic [67]. The PEG-modified nanoparticles can have high solubility in cell membranes which facilities their transition across the cell membrane [68]. However, the PEG modified nanoparticles do not have the functional groups to bind to the target cells. Therefore, the PEG moiety is modified with functional silanes containing amine and catboxylic acid groups which can then bind to many target cells, such as tumor cells. The mechanism by which magnetic nanoparticles are modified by PEG-silane to enable them to conjugate with the target cell is shown in Fig. 7. Studies [69, 70] reported that the uptake of iron magnetic nanoparticles modified by PEG-silane into the macrophage cells was significantly lower than the uptake of unmodified nanoparticles. This observation could be because PEG-modified nano-particles resisted to the adsorption of proteins from the culture medium. Moreover, the uptake of iron magnetic nano-particles modified by PEG-silane into the target cells was significantly higher than the uptake of unmodified nanoparticles. The increase in uptake is probably due to the higher solubility of PEG-modified nanoparticles.

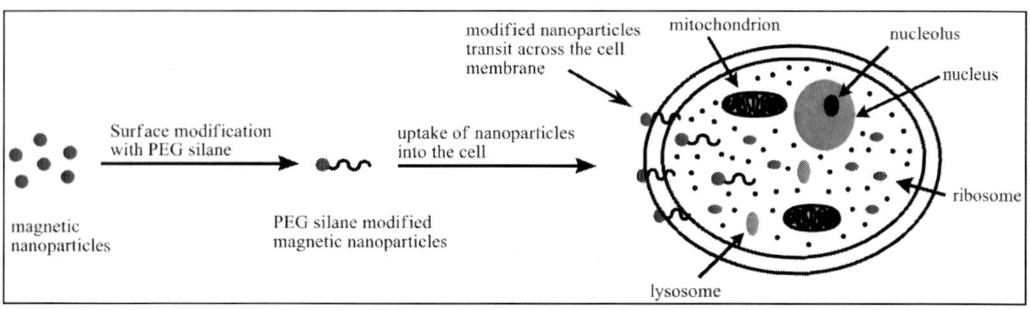

Figure 7. Uptake of surface-modified magnetic nanoparticles into the target cell [modified from ref. 70].

4.2.2.3. DNA Microarrays

The DNA microarray technique is used in disease screening, gene expression analysis and nucleic acid sequencing [71]. There are two methods for binding DNA onto the solid phase of a microarray. The first method is the direct synthesis of DNA binding sites on the solid surface [72]. The second method is the surface modification of the solid support by coating it with a thin layer of silane coupling agents [73]. Two silanes are used in a DNA microarray: i. (3-glycidyloxypropyl)trimethoxysilane, and ii. (N,N-diethyl-3-aminopropyl) trimethoxysilane. The epoxy group of (3-glycidyloxypropyl)trimethoxysilane reacts with the amino group of the DNA oligonucleotide to form a covalent linkage. For (N,N-diethyl-3-aminopropyl)trimethoxysilane, a cross-linking functional agent is added to connect this silane indirectly to the DNA oligonucleotide to form the linkage indirectly. Finally, fluorescent-labeled molecules are incorporated into the assay. After the hybridization process, the assay sample is rinsed to remove unattached fluorescent-labeled molecules. Attached with fluorescent-labeled molecules will emit fluorescent signals to indicate the expressed genes in the samples.

4.2.3. Latest Studies

In DNA microarrays, it is desirable to have strong fluorescent signals for detection. Bifunctional silanes with three hydrolyzable alkoxy groups are always used. A recent study shows that such bifunctional silanes with two hydrolysable alkoxy groups would enhance the florescent signal: Solomum *et al.* [74] explained that the functional silanes with two hydrolysable alkoxy groups form shorter linear siloxane chains than the functional silanes with three hydrolysable alkoxy groups. The formation of such short siloxane chains permits a DNA conformation that extends further away from the surface which is more favorable for reaction with the dye-labeled oligonucleotide. Further studies on the effects of silane tethers on the DNA hybridization are in progress.

4.3. Industry

4.3.1. Fiber-Reinforced Resin Composites

Fiber-reinforced resin composites have many applications in industry. Examples are in materials used in spaceships, automotive vehicles, aircraft, construction and electronics [75]. Different types of fiber are used for reinforcement of the resin matrix. They can be categorized into natural and man-made fibers. Natural fibers come from animals, plant and minerals. Examples are cotton, cellulose, and wood. Examples of man-made fibers are such as E-glass, carbon. Man-made fibers, in general, have better mechanical properties but are more expensive [76].

Silane coupling agents are used to enhance of the bonding at the interfacial fiber-resin matrix [77]. The hydrolyzable alkoxy group of silane can bond to the surface hydroxyl groups of the fiber. The functional group in the other end of the silane can react with the functional group of resin monomers. The silanized fiber is more resistant againt water leaching [78]. Furthermore, surface modification of fibers with hydrophobic silanes enables the reinforcement of non-polar resin matrices. This effect could be due to the reduction in the number of surface hydroxyl groups on the fibers after silanization; hence agglomeration of fibers is reduced during mixing with non-polar resin matrices [79]. The thermal stability of silane modified fibers can also be enhanced which is advantageous because the processing of resin composite requires heat treatment. An enhancement of thermal stability of the fiber means that it would not decompose during heat processing [80]. There is deteriorated effect on the fiber tensile strength when increasing acid concentration for treatment of fiber surfaces before silanization. This could be due to ion exchange reactions at the glass surface by the hydrogen ions and it thus accounts for the reduction in the tensile strength of the composites [81].

4.3.2. Corrosion Protection

Metals or metal alloys are readily oxidized in moist air and thus a thin layer of oxide is formed on the surface. The presence of this oxide layer will change the physical and chemical properties of the metal or metal alloys and its original function is lost. To minimize the corrosion, a protective coating is applied onto the surface. Chromate or phosphate compounds which were efficient corrosion inhibitors were used to coat the metals and metal alloys to prevent corrosion. However, chromates and other chromium compounds have been reported to be toxic and carcinogenic [82]. For this reason, many compounds have been investigated as alternatives to replace chromates. Silanes have been found to be very promising for corrosion protection as they are non-toxic and environmentally clean [83] (Fig. 8).

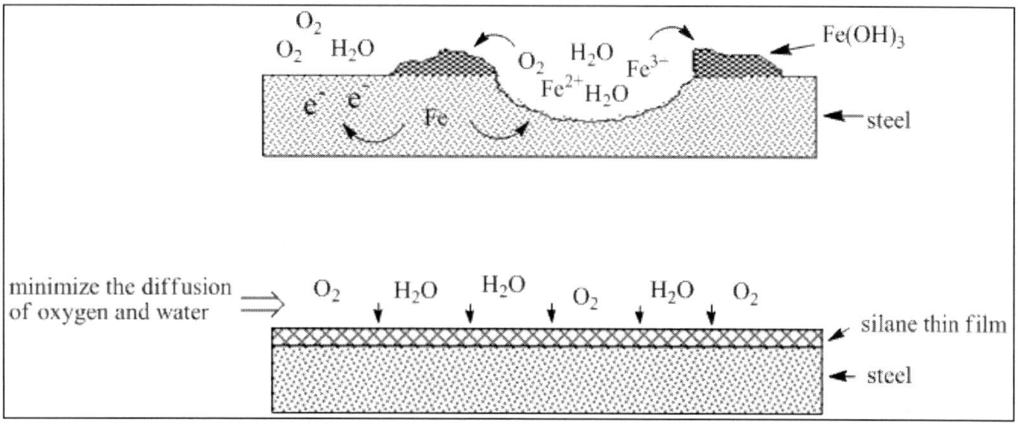

Figure 8 Corrosion mechanism and corrosion protection by a silane coating on steel.

Silanes can be applied onto the metal or metal alloys surfaces by spraying, immersion, rolling, electro-deposition, or brushing. Among these methods, electro-deposition of silane coating gives a more organized, uniform film that is free of voids. Other methods produce non-uniform and uncontrolled thickness [84].

The efficiency of silane coating for corrosion protection is evaluated by corrosion tests such as the atmospheric exposure test, salt spray test, and electrochemical impedance spectroscopy test [85]. Some studies [86-88] reported that silanes such as 3-aminopropyltriethoxysilane, 3-glycidoxypropyltrimethoxysilane, ureidopropyltriethoxysilane, vinyltrimethoxysilane, and functional silanes with cross-linking silanes added such as bis-[trimethoxysilylpropyl]amine performed well in corrosion tests.

4.3.3. Nano-Imprint Lithography

Nano-imprint lithography is a technique of fabricating for nanometer-scale patterns. In this process, a thin resist of thermoplastic polymer is cast onto the substrate by curing. A mold with the desired pattern is pressed into this thin resist cast and cured at an adequate temperature and pressure, thereby creating a thickness contract pattern in the resist. When the mold is removed, the remains of the resist in the compressed areas are removed by reactive ion etching (Fig. 9) [89].

To obtain good contract patterns, the adhesion between the mold and the thermoplastic resist is weak such that the amount of resist residue remaining in the mold after imprinting is small. It is therefore important that the surfaces of the mold and substrate have good anti-adhesive properties, as provided by coating with silanes. This coating improves the imprint qualities and also increases the lifetime of the mold and substrate [90]. Two methods are employed for silane coating: liquid phase deposition and chemical vapour deposition [91, 92].

These studies have investigated the silane coating using fluorinated silanes. The anti-adhesive properties of the mold surfaces were found to be improved.

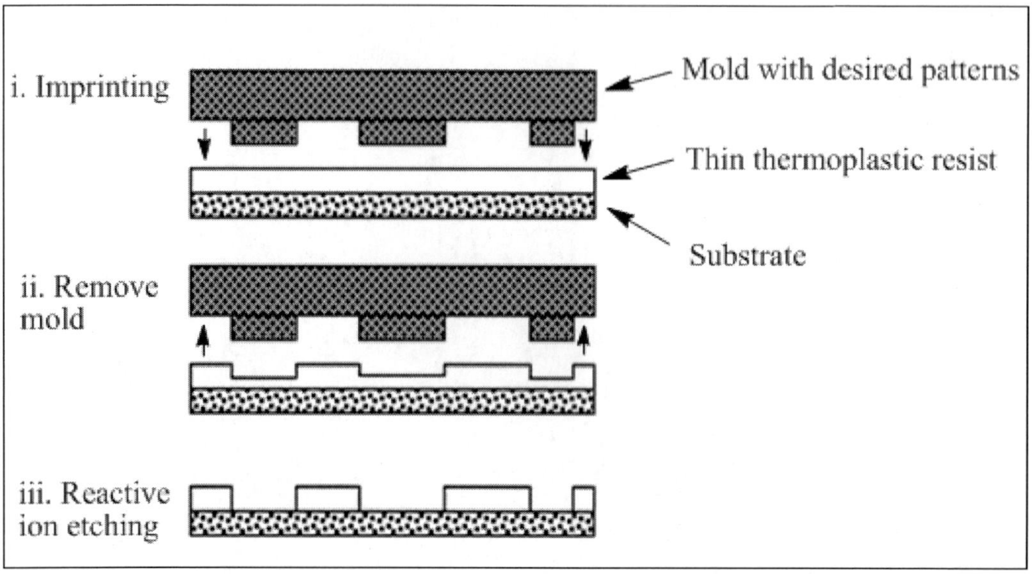

Figure 9. Process of nano-imprint lithography.

4.3.4. Other Applications

Bis-functional silanes such as bis(triethoxysilylpropyl)tetrasulfide (TESPT) and bis(triethoxysilylpropyl)disulfide (TESPD) are used to reinforce of silica-filled elastomer. The bis-silane binds to the silica fillers through the silanol groups by hydrolysis, and to the elastomer by the crosslinking of sulphur atoms with the >C=C< in the elastomer [93]. This binding results in the improvement of the mechanical properties and the processability of the elastomer [94, 95]. A bis-silane modified silica-elastomer has been used in the manufacture of various products, including car tires, gas masks, protective gloves, *etc.* [95].

In a liquid crystal display (LCD) device, the alignment of the nematic liquid crystals is a critical factor to produce sharp, clear and persistent images. A silane coating can enhance the homeotropic orientation of the nematic liquid crystals [96]. The silane used is hydrophobic in nature, *e.g.* a long back-bone chain chlorodimethyloctadecyl silane, $CH_3(CH_2)_{17}Si(CH_3)_2Cl$. The homeotropic alignment of the nematic liquid crystals is caused by weak anchoring and is enhanced when the surface is modified by the hydrophobic silane coating [97, 98].

Silanes are also used as the bonded stationary phase in chromatographic separation systems such as for high pressure liquid chromatography, gas chromatography and thin-layer chromatography. The silanol groups of silanes are bonded to the silica support of the column. By varying the nature of the other functional groups of the silane, the column can have different binding affinities to various substrates from inorganic metal ions to organic compounds [99,100]. Thus, when a mixture of the compounds is carried by with a mobile phase, either liquid or gas, through the column, each component is separated because the retention time is different depending on the strength of the interaction with the stationary phase.

4.3.5. Current Investigations

Recent development in the use of silanes to improve corrosion protection falls into three areas: (1) corrosion resistance with improvement in mechanical properties of silane-coated substrate surfaces by adding filler particles, (2) coating of resin-silane systems to enhance corrosion protection and paint adhesion of metals, and (3) "self-healing" silane coating systems [101].

Suegama *et al.* [102] reported that filling a silane coating film with silica nanoparticles can improve its anti-corrosion behavior and mechanical properties. Silane films contain small pores or micro-cracks in the silane films such that water, oxygen and electrolytes can diffuse and reach the interfacial layer, thus allowing corrosion to still occur. The addition of the silica nanoparticles could enhance the barrier properties of the silane coating. The effect of epoxy incorporated with glycidoxypropyltrimethoxysilane, vinyltrimethoxysilane or dodecyl-trimethoxysilane on corrosion protection of aluminium alloy has been investigated [103]. The results showed that corrosion protection varied for the different silanes, but all of them showed improvement over epoxy resin coating alone. The best performance was observed for the epoxy that was incoraporated with glycidoxypropyltrimethoxysilane.

A "self-healing" silane system is defined as when a silane-coated substrate surface that, with a trace amount of a corrosion inhibitor added, when damaged owing to pinpoints or scratches, reforms a protective layer [104]. This property can prolong the corrosion protection of substrates. A recent study [105] reported that adding corrosion inhibitors such as Ce^{3+} into the silane matrix conferred "self-healing" ability. The mechanism of Ce^{3+} inhibition is through the OH^- generated during corrosion, which reacts with the Ce^{3+} to form insoluble cerium hydroxide/oxide that then becomes deposited on cracks or defects. The damaged silane coating is thus "healed".

5. FUTURE TRENDS OF SILANES

Current silane applications in medicine, dentistry and industry have been discussed. Future developments for silanes include the improvement of bond hydrolytic stability at the interfacial layer and the improvement of adhesion on some inert materials such as graphite, gypsum and zirconia. These developments could solve current problems in medical and dental applications-namely the degradation of the siloxane bond linkage in contact with human body fluid such as in the oral environment. One approach could be to use certain oligomeric silanes that contain a long backbone hydrophobic hydrocarbon chain. The hydrophobic siloxane network formed by these silane molecules reduces the interaction with water molecules. As a result, this increases the bond hydrolytic stability.

The surfaces of inert materials such as carbon blacks and graphite contain with very low concentration of surface hydroxyl groups and would not react with silanes. One approach to enhance the adhesion between silanes and these materials is by activation through oxidation using strong oxidizing agent e.g. potassium permanganate. After oxidation, functional groups of –OH and –COOH are formed on the surfaces which then could be grafted by silane coupling agents.

CONCLUSION

Silanes are widely used in many areas: from biomaterials science to industrial-scale applications. Nowadays, silanes are commonly used even though there are other classes of coupling agents such as phosphates becoming available to serve as alternatives in surface modification and adhesion promotion. The other classes, however, cannot replace silanes completely owing to silanes' many advantages: they are non-toxic and inexpensive, and they display good performance in adhesion promotion. Silanes are still playing and are going to play an important role in biomaterials science.

REFERENCES

[1] Plueddeman EP, *Silane coupling agents. 2nd edition*. 1992; pp.242-7. Plenum Press, New York.
[2] Arkles B. *Silane coupling agents: connecting across boundaries*. 2006; pp. 2-20. Gelest, Inc.
[3] Plueddemann EP. Adhesion through silane coupling agents. *J Adhes* 1970; 2: 184-201.
[4] Plueddemann EP. Water is the key to new theory on resin-to-fiber bonding. *Mod Plast* 1970; 3: 92-8.
[5] Bürger H. Anomalies in the structural chemistry of silicon. *Angew Chem Int Ed Engl* 1973; 12: 474-86.
[6] Heslop RB, Robinson PL. *Inorganic Chemistry – A guide to advanced study. 3rd ed.* Amsterdam, Elsevier. 1967, pp. 363-78.
[7] Gilman H, Dunn GE. Relationship between analogous organic compounds of silicon and carbon. *Chem Rev* 1953; 52: 77-115.
[8] West R. Chemistry of the silicon-silicon double bond. *Angew Chem Int Ed Engl* 1987; 26: 1201-11.
[9] Arkles B. Silicon esters. In: *Kirk-Othmer Encyclopedia of Chemical Technology*. K Othmer editor. London: John Wiley & Sons. 1997; p. 69-81.
[10] West R, Barton TJ. Organosilicon Chemistry. *J Chem Educ* 1980; 57: 165-9.
[11] Deschler U, Kleinschmit P, Panster P. 3-Chloropropyltrialkoxysilanes – key intermediates for the commercial production of organofunctionalized silanes and polysiloxanes. *Angew Chem Int Ed Engl* 1986; 25: 236-52.
[12] Elschenbroich Ch, Salzer A. *Organometallics – A concise introduction. 2ed edition.* 1992; p.2. VCH, Weinhein.
[13] Chambers RC, Jones WE, Haruvy Y, Webber SE, Fox MA. Influence of steric effects on the kinetics of ethyltrimethoxysilane hydrolysis in a fast sol-gel system. *Chem Mater* 1993; 5: 1481-6.
[14] Lung CYK, Matinlinna JP. Resin Bonding to Silicatized Zirconia with Two Isocyanatosilanes and a Cross-linking Silane. *Part II: Mechanistic Approach. Silicon* 2010; 2: 163-9.
[15] Lung CYK, Matinlinna JP, Kukk E, Hägerth T. Surface modification of zirconia by various chemical treatments. *Appl Surf Sci*, 2010; 257: 1228-35.

[16] Matinlinna JP, Özcan M, Lassila LVJ, Vallittu PK. Applications of trialkoxysilanes in dental biomaterials: A review. *Silanes and other coupling agents, vol. 4.* VSP, 2007; 199-219.

[17] Matinlinna JP, Vallittu PK. Silane based concepts on bonding resin composite to metals. *J Cont Dent Pract* 2007; 8: 1-8.

[18] Liu D, Matinlinna JP, Pow EHN. Insights into porcelain zirconia bonding. *J Adhes Sci Technol* 2012; 26: 1249-65.

[19] Guo CY, Tang ATH, Matinlinna JP. Insights into surface treatment methods of titanium dental implants. *J Adhes Sci Technol* 2012; 26: 189-205.

[20] Ho GW, Matinlinna JP. Insights on porcelain as a dental material. Part II: Chemical surface treatments. *Silicon* 2011; 3: 117-23.

[21] Matinlinna JP, Vallittu PK. Bonding of resin composites to etchable ceramic surfaces – an insight review to the chemical aspects on surface conditioning. *J Oral Rehabil* 2007; 34: 622-30.

[22] Heikkinen TT, Lassila LVJ, Matinlinna JP, Vallittu PK. Effect of operating air pressure on tribochemical silica-coating. *Acta Odontol Scand* 2007; 65: 241-8.

[23] Lung CYK, Matinlinna JP, Resin bonding to silicatized zirconia with two isocyanatosilanes and a cross-linking silane. *Part I Experimental. Silicon* 2010; 2: 153-61.

[24] Gbureck U, Masten A, Probst J, Thull R. Tribochemical structuring and coating of implant metal surfaces with titanium oxide and hydroxyapatite layers. *Mater Sci Eng* 2003; 23: 461-5.

[25] Chaiyabutr Y, McGowan S, Philips KM, Kois JC, Giordano RA. The effect of hydrofluoric acid surface treatment and bond strength of a zirconia veneering ceramic. *J Prosthet Dent* 2008; 100: 194-202.

[26] Ö zcan M, Vallittu PK. Effect of surface conditioning methods on the bond strength of luting cement to ceramics. *Dent Mater* 2003; 19: 725-31.

[27] Janda R, Franç ois JF, Wulf M, Tiller HJ. Resin/resin bonding: a new adhesive technology. *J Adhes Dent* 2002; 4: 299-308.

[28] Matinlinna JP, Lassila LVJ, Özcan M, Yli-Urpo A, Vallittu PK. An introduction to silanes and their applications in dentistry. *Int J Prosthodont* 2004; 17: 155-64.

[29] Alex G. Preparing porcelain surfaces for optimal bonding. *Comp Cont Educ Dent,* 2008; 29: 324-35.

[30] Lee SY, Vang MS, Yang HS, Park SW, Park HO, Lim HP. Shear bond strength of composite resin to titanium according to various surface treatments. *J Adv Prosthodont* 2009; 1: 68-74.

[31] Kern M, Thompson VP. Durability of resin bonds to a cobalt-chromium alloy. *J Dent,* 1995; 23: 47-54.

[32] Xible AA, de Jesus Tavarez RJ, Araujo C, Bonachela W. Effect of silica coating and silanization on flexural and composite resin bond strengths of zirconia posts: an in vitro study. *J Prosthet Dent* 2006; 95: 224-9.

[33] Lung CYK, Matinlinna JP. Aspects of silane coupling agents and surface conditioning in dentistry – An overview. *Dent Mater* 2012; 28: 467-77.

[34] Eichmiller FC. Titanium applications in dentistry. *J Am Dent Assoc* 2003; 134: 347-9.

[35] Sharma M, Kumar AVR, Singh N, Adya N, Salujia B. Electrochemical corrosion behavior of dental/implant alloys in artifical saliva. *JMEPEG* 2008; 17: 695-701.

[36] Shimoe S, Tanoue N, Yanagida H, Atsuta M, Koizumi H, Matsumura H. Comparative strength of metal-ceramic and metal-composite bonds after extended thermocycling. *J Oral Rehabil*, 2004; 31: 689-94.

[37] Lee JH, Um CM, Lee IB. Rheological properties of resin composites according to variations in monomer and filler composition. *Dent Mater* 2006; 22: 515-26.

[38] Drummond JL. Degradation, fatigue and failure of resin dental composite materials. *J Dent Res* 2008; 87: 710-9.

[39] Lin CT, Lee SY, Keh ES, Dong DR, Huang HM, Shih YH. Influence of silanization and filler fraction on aged dental composites. *J Oral Rehabil* 2008; 27: 919-26.

[40] Zhang M, Matinlinna JP. The effect of resin matrix composition on mechanical properties of E-glass fiber reinforced composite for dental use. *J Adhes Sci Techn* 2011; 25: 2687-701.

[41] Aboushelib MN, Kleverlaan CJ, Feilzer AJ. Selective infiltration-etching technique for a strong and durable bond of resin cements to zirconia-based materials. *J Prosthet Dent* 2007; 98: 379-88.

[42] Akyil MŞ, Uzun IH, Bayindir F. Bond strength of resin cement to yttrium-stabilized tetragonal zirconia ceramic treated with air abrasion, silica coating and laser irradiation. *Photomed Laser Surg* 2010; 28: 801-8.

[43] Piascik JR, Swift EJ, Thompson JY, Grego S, Stoner BR. Surface modification for enhanced silanation of zirconia ceramics. *Dent Mater* 2009; 25: 1116-21.

[44] Matinlinna JP, Lassila LV. Experimental novel silane system in adhesion promotion between dental resin and pretreated titanium. *Part II: Effect of long term water storage. Silicon* 2010; 2: 79-85.

[45] Matinlinna JP, Heikkinen T, Özcan M, Lassila LVJ, Vallittu PK. Evaluation of resin adhesion to zirconia ceramic using some organosilanes. *Dent Mater* 2006; 22: 824-831.

[46] Lung CYK, Botelho MG, Heinonen M, Matinlinna JP. Resin zirconia bonding promotion with some novel coupling agents. *Dent Mater* 2012; doi: 10.1016/j.dental.2012.04.023.

[47] Matinlinna JP, Lassila LV. Enhanced resin-composite bonding to zirconia framework after pretreatment with selected silane monomers. *Dent Mater* 2011; 27: 273-280.

[48] Matinlinna JP, Lassila LVJ, Kangasniemi I, Vallittu PK. Isocyanato- and methacryloxysilanes promote bis-GMA adhesion to titanium. *J Dent Res* 2005; 84: 360-64.

[49] Matinlinna JP, Lassila LVJ, Vallittu PK. Pilot evaluation of resin composite cement adhesion to zirconia using a novel silane system. *Acta Odontol Scand* 2007; 65: 44-51.

[50] Matinlinna JP, Özcan M, Lassila LVJ, Vallittu PK. The effect of a 3-methacryloxypropyltrimethoxysilane and vinyltriisopropoxysilane blend and *tris*(3-trimethoxysilylpropyl)isocyanurate on the shear bond strength of composite resin to titanium metal. *Dent Mater* 2004; 20: 804-13.

[51] Matinlinna JP, Lassila LVJ, Vallittu PK. The effect of five silane coupling agents on the bond strength of a luting cement to a silica-coated titanium. *Dent Mater* 2007; 23: 1173-80.

[52] Matinlinna JP, Lassila LVP, VallIttu PK. The effect of three silane coupling agents and their blends with a cross-linker silane on bonding a bis-GMA resin to silicatized titanium (A novel silane system). *J Dent* 2006; 34: 740-6.

[53] Lung CYK, Kukk E, Matinlinna JP. Shear bond strength between resin and zirconia with two different silane blends. *Acta Odontol Scand* 2012; doi: 10.3109/00016357.2011.630014.

[54] Matinlinna JP, Özcan M, Lassila L, Kalk W, Vallittu PK. Effect of the cross-linking silane concentration in a novel silane system on bonding resin-composite cement. *Acta Odontol Scand* 2008; 66: 250-5.

[55] Erli HJ, Marx R, Paar O, Niethard FU, Weber M, Wirtz DC. Surface pretreatments for medical application of adhesion. *BioMed Eng Online* 2003; 2: 8 pages.

[56] Mumme T, Marx R, Müller-Rath R, Siebert CH, Wirtz DC. Surface coating to improve the metal-cement bonding in cemented femur stems. *Arch Orthop Trauma Surg* 2008; 128: 773-81.

[57] Müller RT, Schürmann N. Shear strength of the cement metal interface: An experimental study. *Arch Orthop Trauma Surg* 1999; 119: 133-8.

[58] Raab S, Ahmed AM, Provan JW. Thin film PMMA precoating for improved implant bone-cement fixation. *J Biomed Mater Res* 1982; 16: 679-704.

[59] Yerby SA, Paal AF, Young PM, Beaupre GS, Ohashi KL, Goodman SB. The effect of a silane coupling agent on the bond strength of bone cement and cobalt-chrome alloy. *J Biomed Mater Res* 2000; 49: 127-33.

[60] Fischer H, Wirtz DC, Weber M, Neuss M, Niethard FU, Marx R. Improvement of the long-term adhesive strength between metal stem and polymethylmethacrylate bone cement by a silica/silane interlayer system. *J Biomed Mater Res* 2001; 57: 413-8.

[61] Pope NM, Kulcinski DL, Hardwick A, Chang YA. New application of silane coupling agents for covalently binding antibodies to glass and cellulose solid supports. *Bioconjugate Chem* 1993; 4: 166-71.

[62] Costerton JW, Stewart PS, Greenberg EP. Bacterial biofilms: a common cause of persistent infections. *Science* 1999; 284: 1318-22.

[63] Stewart PS, Costerton JW. Antibiotic resistance of bacteria in biofilms. *Lancet* 2001; 358: 135-8.

[64] Mohorčič M, Jerman I, Zorko M, Butinar L, Orel B. Jerala R, Friedrich J. Surface with antimicrobial activity obtained through silane coating with covalently bound polymyxin B. *J Mater Sci: Mater Med* 2010; 21: 2775-82.

[65] Bergemann C, Müller-Schulte D, Oster J, à Brassard L, Lübbe AS. Magnetic ion-exchange nano- and microparticles for medical, biochemical and molecular biological applications. *J Magn Magn Mater* 1999; 194: 45-52.

[66] Allemann E, Gurny R, Doelker E. Drug-loaded nanoparticles – preparation methods and drug targeting issues. *Eur J Pharm* 1993; 39: 173-91.

[67] Kohler N, Fryxell GE, Zhang M. A bifunctional poly(ethylene glycol) silane immobilized on metallic oxide-based nanoparticles for conjugation with cell targeting agents. *JACS* 2004; 126: 7206-11.

[68] Boni LT, Hah JS, Hui SW, Mukherjee R, Ho JT, Jung CY. Aggregation and fusion of unilamellar vesicles by poly(ethylene glycol). *Biochem Biophys Acta* 1984; 775: 409-18.

[69] Zhang Y, Sun C, Kohler N, Zhang M. Self-assembled coatings on individual monodisperse magnetite nanoparticles for efficient intracellular uptake. *Biomed Microdevices* 2004; 1: 33-40.

[70] Zhang Y, Kohler N, Zhang M. Surface modification of superparamagnetic magnetite nanoparticles and their intracellular uptake. *Biomater* 2002; 23: 1553-61.

[71] Ramsay G. DNA chips: state-of-the art. Nat Biotechnol 1998; 16: 40-4.

[72] Cohen G, Deutsch J, Fineberg J, Levine A. Covalent attachment of DNA oligonucleotides to glass. *Nucleic Acids Res* 1997; 25: 911-2.

[73] Saal K, Tätte T, Tulp I, Kink I, Kurg A, Mäeorg U, Rinken A, Lõhmus A. Sol-gel films for DNA microarray applications. *Mater Lett* 2006; 60: 1833-8.

[74] Solomum T, Mix R, Sturm H. Immobilization of silanized DNA on glass: influence of the silane tether on the DNA hybridization. *ACS Appl Mater Interfaces* 2010; 2: 2171-4.

[75] Zhang M, Matinlinna JP. E-glass fiber reinforced composites in dental applications. *Silicon* 2012; 4: 73-8.

[76] Bledzki AK, Gassan J. Composites reinforced with cellulose based fibers. *Prog Polym Sci* 1999; 24: 221-74.

[77] Plueddemann EP, Clark HA, Nelson LE, Hofmann KR. New silane coupling agents for reinforced plastics. *Mod Plast* 1962; 39: 135-87.

[78] Abdelmouleh M, Boufi S, ben Salah A, Belgacem MN, Gandini A. Interaction of silane coupling agents with cellulose. *Langmuir* 2002; 18: 3203-8.

[79] Donath S, Militz H, Mai C. Creating water-repellent effects on wood by treatment with silanes. *Holzforschung* 2006; 60: 40-6.

[80] Arbelaiz A, Fernández B, Ramos JA, Mondragon I. Thermal and crystallization studies of short flax fibre reinforced polypropylene matrix composites: effect of treatments. *Thermochim Acta* 2006; 440: 111-21.

[81] Sever K, Sarikanat M, Seki Y, Cecen V, Tavman IH. Effects of fiber surface treatments on mechanical properties of epoxy composites reinforced with glass fabric. *J Mater Sci* 2008; 43: 4666-72.

[82] van Ooij WJ, Zhu D, Prasad S, Jayaseelan S, Fu Y, Teredesai N. Silane based chromate replacements for corrosion control, paint adhesion and rubber bonding. *Surf Eng* 2000; 16: 386-96.

[83] van Ooij WJ, Child T. Protecting metals with silane coupling agents. Chemtech 1998; 28: 26-35.

[84] Gandhl JS, van Ooij WJ. Improved corrosion protection of aluminium alloys by electrodeposited silanes. *JMEPEG* 2004; 13: 475-80.

[85] Chico B, de la Fuente D, Pérez ML. Corrosion resistance of steel treated with different silane/paint systems. *J Coat Technol Res* 2012; 9: 3-13.

[86] Fedel M, Olivier M, Poelman M, Deflorian F, Rossi S, Druart ME. Corrosion protection properties of silane pre-treated powder coated galvanized steel. *Prog Org Coat* 2009; 66: 118-28.

[87] Zhu D, van Ooij WJ. Corrosion protection of metals by water-based silane mixtures of bis-[trimethoxysilylpropyl]amine and vinyltriacetoxysilane. *Prog Org Coat* 2004; 49: 42-53.

[88] Puomi P, Fagerholm HM. Performance of silane treated primed hot-dip galvanised steel. *Anti-Corros Method M* 2001; 48: 7-17.

[89] Chou SY, Krauss PR, Renstrom PJ. Imprint lithography with 25-nanometer resolution. *Science* 1996; 272: 85-7.

[90] Park S, Schift H, Padeste C, Schnyder B, Kötz R, Gobrecht J. Anti-adhesive layers on nickel stamps for nanoimprint lithography. *Microelectron Eng* 2004; 73-74: 196-201.

[91] Beck M, Graczyk M, Maximov I, Sarwe EL, Ling TGI, Keil M, Montelius L. Improving stamps for 10 nm level wafer scale nanoimprint lithography. *Microelectron Eng* 2002; 61-62: 441-8.

[92] Li JR, Lusker KL, Yu JJ, Garno JC. Engineering the spatial selectivity of surfaces at the nanoscale using particle lithography combined with vapour deposition of organosilanes. *ACSNano* 2009; 3: 2023-35.

[93] ten Brinke JW, Debnath SC, Reuvekamp LAEM, Noordermeer JWM. Mechanistic aspects of the role of coupling agents in silica-rubber composites. *Compos Sci Technol* 2003; 63: 1165-74.

[94] Ismail H, Kheong OW. The effect of bis-(3-triethoxysilylpropy) – tetrasulfide, Si69 on properties of recycled poly(vinyl chloride)/acrylonitrile – butadiene rubber/fly ash (PVCr/NBR/FA) composites. *J Reinf Plast Comp* 2008; 27: 1649-61.

[95] Kim KJ. Bifunctional silane (TESPD) effects on improved mechanical properties of silica containing nitrile-butadiene rubber/poly(vinyl chloride) compound. *J Appl Polym Sci* 2012; 124: 2937-44.

[96] Abderrahmen A, Romdhane FF, Ouada HB, Gharbi A. Investigation of the liquid crystal alignment layer: effect on electrical properties. *Sci Technol Adv Mate*r 2008; 9: 1-5.

[97] Cho YK, Granick S. A surface forces platform for dielectric measurements. *J Chem Phys* 2003; 119: 547-54.

[98] Akiyama H, Iimura Y. Hometropic alignment properties of nematic liquid crystal on uv-exposed alignment film. *Jpn J Appl Phys* 2001; 40: L765-8.

[99] Kirkland JJ. Bidentate silane stationary phases for reversed-phase high-performance liquid chromatography. *Anal Chem* 1998; 70: 4344-52.

[100] Henry JB, Sweet TR. Chemically-bonded stationary phases for thin-layer chromatography of metal ions. *Chromatographia* 1983; 17: 79-82.

[101] van Ooij WJ, Zhu D, Palanivel V, Lamar A, Stacy M. Overview: The potential of silanes for chromate replacement in metal finishing industries. *Silicon Chem* 2006; 3: 11-30.

[102] Suegama PH, de Melo HG, Recco AAC, Tschiptschin AP, Aoki IV. Corrosion behaviour of carbon steel protected with single and bi-layer of silane films filled with silica nanoparticles. *Surf Coat Tech* 2008; 202: 2850-8.

[103] Ji WG, Hu JM, Liu L, Zhang JQ, Cao CN. Improving the corrosion performance of epoxy coatings by chemical modification with silane monomers. *Surf Coat Tech* 2007; 201: 4789-95.

[104] Moutarlier V, Neveu B, Gigandet MP. Evolution of corrosion protection for sol-gel coatings doped with inorganic inhibitors. *Surf Coat Tech* 2008; 202: 2052-8.

[105] Luo F, Li Q, Zhong XK, Gao H, Dai Y, Chen FN. Corrosion electrochemical behaviors of silane coating coated magnesium alloy in NaCl solution containing cerium nitrate. *Mater Corros* 2012; 63: 148-54.

In: Silane: Chemistry, Applications and Performance
Editors: Katsuo Moriguchi and Susumu Utagawa
ISBN: 978-1-62257-432-2

Chapter 4

CATIONIC RING OPENING POLYMERIZATION OF ε-CAPROLACTONE USING SILANES: TUNING OF MOLECULAR WEIGHT AND IN-SITU INCORPORATION OF NANOPARTICLES

Mohamad-Ali Tehfe[1,*], Rachid Souane[1], Jacques Lalevée[1], Mathieu Peter[1], Didier Gigmes[2], Laurence Charles[2], Didier Le Nouen[3] and Jean Pierre Fouassier[4]

[1]Institut de Science des Matériaux de Mulhouse, Mulhouse, France
[2]Institut de Chimie Radicalaire ICR, Aix-Marseille Université, Marseille, France
[3]Laboratory of Organic and Bioorganic Chemistry, University of Haute Alsace, Mulhouse, France
[4]UHA-ENSCMu, Mulhouse, France

ABSTRACT

New initiating systems based on silanes and metal salts are proposed for an easy thermal cationic polymerization of ε-caprolactone at RT under air. They ensure good to excellent polymerizations. Moreover, a direct incorporation of the metal (Ag, Pd) through this one pot synthetic procedure is observed. The addition of small amounts of water or isopropanol allows a tuning of the polycaprolactone molecular weights. A coherent picture of the involved chemical mechanisms is presented.

Keywords: Silanes, Silver Salts, Cationic Polymerization, Polycaprolactone, Epoxides

[*] Author E-mail address: mohamad-ali.tehfe@uha.fr, j.lalevee@uha.fr.

INTRODUCTION

Cationic polymerization (CP) reactions are very attractive. [1] CP has been successfully applied to a large variety of monomers such as cyclic ethers and esters and related derivatives, epoxides, vinyl ethers... The mechanical properties of the final material will be different than those of the well known acrylic polymers prepared through free radical polymerization FRP. The use monomers less toxic than the acrylates can also be highly worthwhile. Interestingly, the oxygen inhibition observed in FRP does not occur. The cationic ring opening polymerization ROP of ε-caprolactone ε-CL has received, however, much less proposals than the other modes of activation. [2-8] Recent attempts include the use of HCl, trifluoromethanesulfonic acids, organomagnesium complexes, yttrium triflate as initiators (see [2,4-8] and references therein). The addition of a little amount of water or the use of an alcohol in the reaction mixture enhances the polymerization rate. [2,8] The development of new and efficient cationic initiating systems for the polymerization of ε-CL at RT and under air is still fascinating due to the large applications of the biodegradable polycaprolactone PCL in packaging, medicine, tissue engeenering…; for example, PCL can be used as an additive to improve the processing of resins and/or their end use properties (e.g. their impact resistance); incorporation of nanoparticles (NP) into PCL could give an added value such as an antibactericide property with Ag. [9] The incorporation of Ag(0) through photopolymerization processes has been elegantly presented in [9c,d].

Among various initiation systems for CP of epoxides, redox couples are interesting [10-12] as exemplified by a recent new tri-component combination [12-13] based on onium salt and organosilane (R_3SiH) as the reducing agent and platinum, palladium or rhodium as a metal catalyst. On the other side, we recently developed a two-component redox system based on a tris(trimethylsilyl)silane TTMSS/silver salt combination which allows both to efficiently initiate the ROP of epoxides at RT and to incorporate metal M(0) nanoparticles into the film (Scheme 1 with M = Ag). [14]

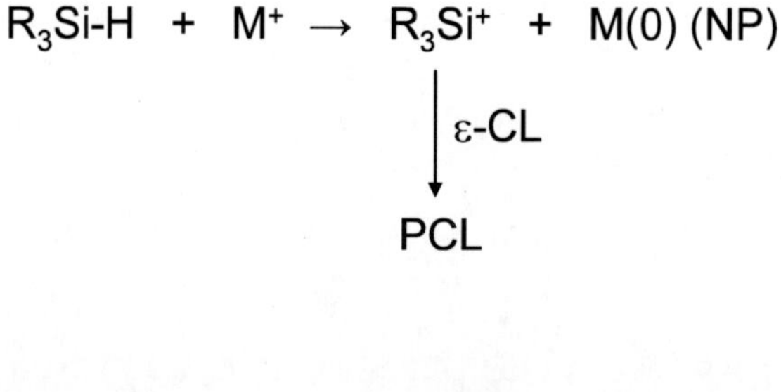

Scheme 1.

In the present chapter, we propose to use such a redox system based on the diphenylsilane DPSi/silver salt couple for the thermal cationic polymerization of ε-caprolactone (Scheme 2). DPSi is preferred here to tris(trimethylsilyl)silane –TTMSS- in order to avoid the side radical reaction route of TTMSS parallel to the redox route. [14] Our system should open up a new way for i) the cationic ROP of ε-CL initiated by a true redox process at RT and under air, ii) the in-situ incorporation of nanoparticles into PCL and iii) a tuning of the molecular weigths through the addition of water or alcohols. The performances of the initiating systems for the comparative ROP of ε-CL and an epoxide EPOX (Scheme 2) as followed by FTIR, NMR and

SEC will be also discussed; mass spectrometry (MALDI-TOF) experiments will help to clarify the initiation mechanism.

EXPERIMENTAL PART

1. Investigated compounds

The investigated compounds (Scheme 2) are used here with the best purity available. Diphenylsilane (DPSi) was selected as a representative silane (Aldrich). Phenyldimethylsilane (PDMSi) from Aldrich and poly(methylhydro)siloxane (PHSi) from Gelest (HMS-991) were also investigated. Silver hexafluoroantimonate ($AgSbF_6$), silver tetrafluoroborate ($AgBF_4$), silver triflate (AgTf), silver acetate (AgOAc), silver chloride (AgCl), palladium (II) chloride ($Pd(Cl)_2$), palladium (II) acetate ($Pd(OAc)_2$) salts were also obtained from Aldrich. (3,4-epoxycyclohexane)methyl 3,4-epoxycyclohexylcarboxylate (EPOX or UVACURE 1500 from Cytec) and ε-caprolactone (ε-CL - Aldrich; used as received) were used as monomers.

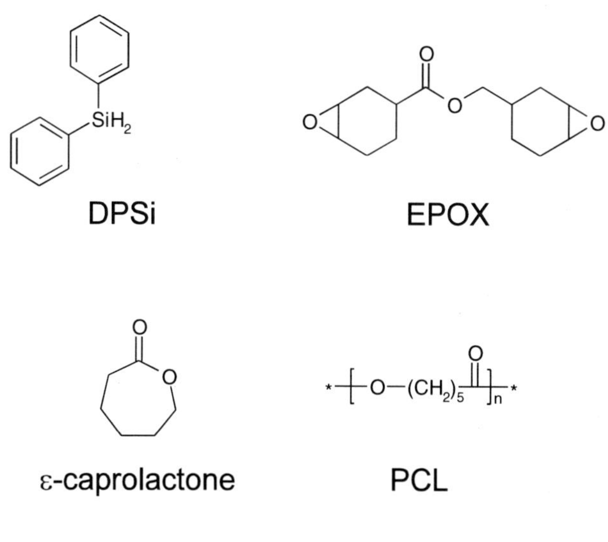

Scheme 2.

2. Cationic Polymerization

The silver or palladium salts (0.5 to 2 % w/w) were dissolved in the selected monomer (EPOX or ε-caprolactone) and diphenylsilane (DPSi) is introduced into the formulation by a syringe at RT.

The polymerization of EPOX or ε-CL is carried out in pill-box for a sample of 1g. A magnetic stirrer is in contact with the pill-box to ensure a good homogeneity of the sample. The addition of the silane into the formulation corresponds to time t = 0 s. For ε-CL, as described in detail in [7], the monomer conversion can be easily followed by NMR as the ratio of the representative peaks of CL vs. those of PCL (see below).

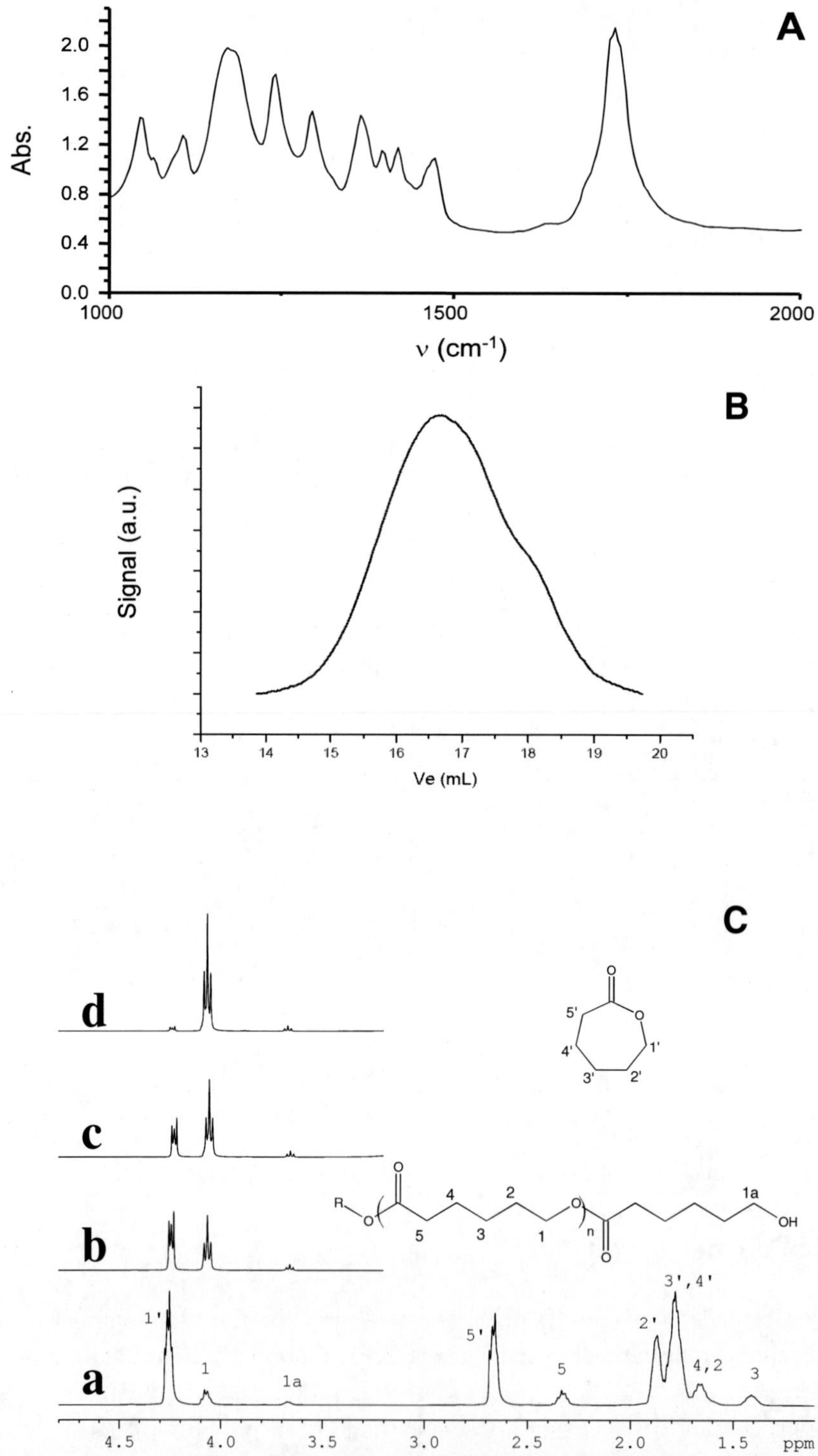

Figure 1. Characterization of the PCL polymer. (A) IR spectrum in KBr; (B) SEC experiments (see text); (C) NMR spectra in $CDCl_3$ characterizing the polymerization of ε-CL at different reaction times: (a) t = 9 min; (b) t = 32 min; (c) t = 52 min; (d) t = 95 min.; initiating system: $Ph_2SiH_2/AgSbF_6/H_2O$ (1%/1%/0.5% w/w). The attribution of the NMR peaks is given.

For EPOX, the progress of the exothermic polymerization is followed by monitoring the sample temperature using a thermocouple connected to a DaqPro-5300 (resolution 0.1°C). The reaction is too slow with ε-CL to use a similar procedure. It has been shown in [12-13] that the increase of the temperature is directly related to the monomer conversion. The present set-up used here being not adiabatic, the relationship between the conversion and the temperature will be valid only for the first steps of the polymerization process.

3. Characterization of the PCL

The IR Spectrum

Transmission FTIR spectrum of PCL dispersed in KBr was done on a Nicolet Nexus 870 apparatus.

The NMR Spectra

^{1}H (400.1MHz) NMR spectra were recorded in $CDCl_3$ at room temperature on a Bruker Avance-400 spectrometer. Chemical shift are given on the δ scale and referenced to tetramethylsilane (TMS) at 0 ppm. The following parameters have been used for the NMR spectra (99.8% EURISO-TOP): acquisition time: 4s, relaxation delay: 10s, pulse duration: 2.1μs for a 30° flip angle and spectra width: 8250.83Hz (20.18 ppm); 64K points were used for the data acquisition; 64K for Fourier transformation and digital resolution: 0.12Hz/point.

Size Exclusion Chromatography (SEC)

The number average molecular weights (M_n) and polydispersity index (PDI=M_w/M_n) were determined by size exclusion chromatography SEC using a Waters 515 HPLC pump equipped with 2 columns (PSS SDV LINEAR M 5 μm), and two detectors: UV/visible (Waters 486) and DRI (Waters 2414). Measurements were performed in tetrahydrofuran (THF) at room temperature, with a flow rate of 1 $mL.min^{-1}$. Toluene was used as a flow rate marker. Calibration was based on polystyrene standards (kit EasyCal from Polymer Laboratories; M_n between 580 and 377,400 $g.mol^{-1}$); Mark-Houwink parameters: for PS: $K = 11.4 \times 10^{-3}$ $mL.g^{-1}$, $\alpha = 0.716$; for PBA: $K = 12.2 \times 10^{-3}$ $mL.g^{-1}$, $\alpha = 0.70$). [15]

Mass Spectrometry

MALDI-TOF MS experiments were carried out using a Bruker Autoflex (Bruker Daltonics, Leipzig, Germany). The instrument is equipped with a nitrogen laser emitting at 337 nm, a single-stage pulsed ion extraction source and dual microchannel plate detectors. Data acquisition was performed in reflectron mode. Positive ion mode was used for all analyses, using a pulse frequency of 10 Hz and an accelerating voltage of 19 kV. The delay time used in delayed extraction mode was optimized based on the mass range of the polymer distribution and was generally around 100 ns. FlexControl software version 2.2 (Bruker Daltonics) was used for instrument control and data acquisition, and FlexAnalysis software version 2.2 (Bruker Daltonics) for data processing. External calibration was performed using poly(propylene glycol). Trans-2-[3-(4-*tert*-butylphenyl)-2-methyl-2-propenylidene]-malonitrile (DCTB) used as the matrix and sodium iodide used as the cationizing agent were from Sigma. Individual solutions of PCL samples, of the matrix (11.7 mg mL^{-1}) and of the salt (8.8 mg mL^{-1}) were prepared in dichloromethane (SDS, Peypin, France). An aliquot (5 μL) of the PCL solution was mixed with 50 μL of the DCTB solution and 2 μL of the NaI solution. MALDI samples were prepared using the dried-droplet method, where 1 μL of the liquid mixture was deposited onto the sample plate and allowed to air-dry.

RESULTS AND DISCUSSION

1. Cationic Polymerization

Using the DPSi/$AgSbF_6$ initiating system (which is the most efficient for the polymerization of EPOX [14]), the polymerization of ε-caprolactone is observed within some hours at RT (the gel time is about 4-6 hours). For DPSi/$AgBF_4$, a similar process occurs but much more slowly (2-3 days at RT). As expected, [14] using high nucleophilic counter-anions (Cl^-, AcO^- ...) does not allow a significant polymerization of ε–CL. The formation of PCL is well evidenced by the NMR and IR spectra of the obtained polymer (Figure 1 A and 1C) which are fully consistent with the already published spectra. [7,16] M_ns and polydispersity indexes (PDI=M_w/M_n) of the polymers obtained were also characterized by size exclusion chromatography SEC (Figure 1B – Table 1). Remarkably, the polymerization is initiated by the new proposed system in simple glass vials, under air and at RT.

A typical ^{1}H NMR series of spectra characterizing the ring opening polymerisation of ε-CL at different reaction times is depicted in Figure 1C. In those spectra, both the consumption of ε-CL and formation of PCL can be followed using the NMR signals representative of the different ^{1}H as usually done. [7] The peak assignment for ^{1}H NMR signal of PCL is fully consistent with that of literature. [16c] The proton signals at 3.66 ppm (H_{1a} methylene end group adjacent to the PCL terminal hydroxyl), 4.06ppm (H_1 core methylene of PCL adjacent to the ester group) and 4.24 ppm ($H_{1'}$ methylene of the CL moiety adjacent to the ester in PCL) are interesting: i) the proton ratio between H_1 and $H_{1'}$ gives an access to the monomer conversion, ii) the proton ratio between H_1 and H_{1a} leads to the degree of polymerization DP and the number average molecular weight $\overline{M_n}$. A typical conversion vs. time plot for ε-CL using DPSi/$AgSbF_6$ is shown in Figure 2A.

Interestingly, the addition of a small amount of water strongly enhances the polymerization rate (Figure 2A). A decrease of the gel time is also noted i.e. ~ 60 and 30 min for 0.1 and 0.5 % w/w of water, respectively. A complete conversion is achieved at ~ 100 min for [H_2O] = 0.5 % w/w (Figure 2A). Since ε-CL is not purified, the presence of an initial amount of H_2O can induce a slight error in the overall H_2O content and can also participate to the initiation process presented above. However, the great interest of the present approach lies on the fact that the new proposed initiating system works using simple glass vials and unpurified ε-CL even in the presence of water or O_2 which can behave in other synthetic approaches as polymerization inhibitors. [8] This can open new opportunities for mass production and applications of PCL. For the other R_3SiH investigated (Phenyldimethylsilane (PDMSi) and poly(methylhydro)siloxane (PHSi)), the combinations with $AgSbF_6$ and water also lead to highly efficient initiating systems demonstrating the versatility of the new proposed approach (gel times of about 30 min). For the addition of isopropanol, the beneficial effect on the polymerization rate is weaker than with water (Figure 2A). A similar result is found using tBuOH instead of isopropanol (*not shown*).

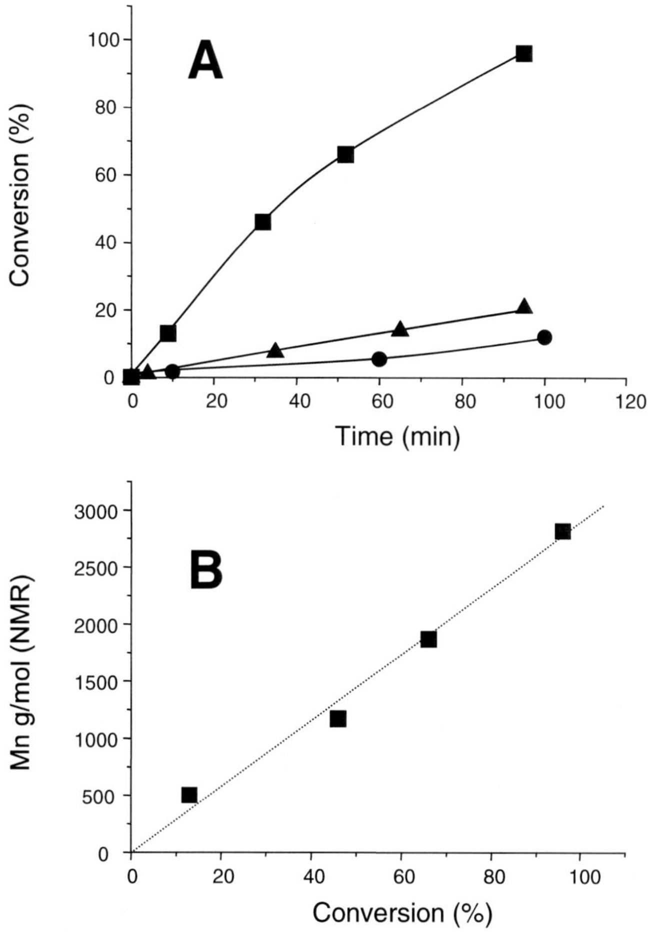

Figure 2. (A) Conversion vs. time curves for the polymerization of ε-CL initiated by DPSi/$AgSbF_6$/H_2O (squares); DPSi/$AgSbF_6$/2-propanol (up triangles) (1%/1%/0.5% w/w) and DPSi/$AgSbF_6$ (1%/1% w/w) (circles). (B) M_n vs. conversion; initiating system: Ph_2SiH_2/$AgSbF_6$/H_2O (1%/1%/0.5% w/w); the M_n are determined by NMR.

A good agreement is obtained between the M_n determined by SEC and NMR (Table 1). Interestingly, the M_n linearly increases with the conversion (Figure 2B and supporting information). The formed PCLs are characterized by a polydispersity index in the 1.5-1.7. This in agreement with previous results on cationic processes. [8] For an initiating system without additional water, the theoretical M_ns (calculated from the degree of polymerization as DP = [monomer]/[initiator]) are in good agreement with the calculated ones (entry 1: 9530 g/mol and entry 2: 13545 g/mol; on the assumption that each Si-H function participate to the initiation of the polymerization). This is in agreement with a decrease of [DPSi] which leads to higher M_n. More interestingly, the addition of water affects the kinetic (see above) but also the M_n values which are noticeably lower. For entries 3-4, the theoretical initiator concentrations deduced from the experimental DPs are about 0.3 M and 0.38 M. These concentrations are much higher than [DPSi] (~ 0.054 M) thereby demonstrating that water also plays a role in the initiating step (such a behavior has been recently observed for the ROP of ε-CL initiated by yttrium triflate [8]). This property is quite useful since the final M_n can be tuned by addition of water (Table 1; entries 5-7).

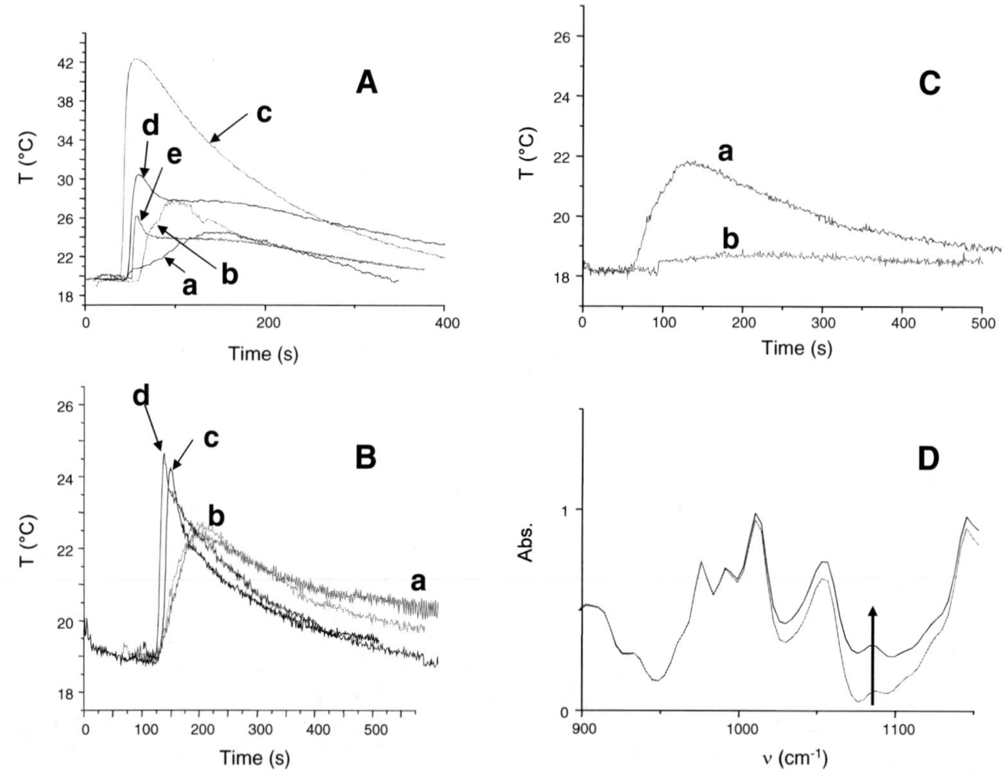

Figure 3. Role of the metal in the cationic polymerization of EPOX. (A) Influence of water: (a) without water; (b) 0.05 % (w/w); (c) 0.1 % (w/w); (d) 0.3 % (w/w); (b) 0.4 % (w/w); DPSi/$AgSbF_6$ (1%/1% w/w) as the initiating system. (B) Comparison of the polymerization initiating ability of palladium salts: (a) DPSi/$Pd(OAc)_2$ (1%/0.5% w/w), (b) DPSi/$Pd(OAc)_2$ (1%/1% w/w), (c) DPSi/$Pd(OAc)_2$ (1%/2% w/w), (d) DPSi/$Pd(OAc)_2$ (1%/4% w/w). (C) Comparison of the polymerization initiating ability of palladium salts vs. silver salts: (a) DPSi/$Pd(OAc)_2$ (1%/1% w/w), (b) DPSi/AgCl (1%/1% w/w). (D) Change of the FTIR spectra upon addition of DPSi (1% w/w) to $Pd(OAc)_2$ (1% w/w).

For comparison, similar observations are also noted when ε-CL is changed for a usual epoxy monomer EPOX (Scheme 2): i) the more efficient process is found with SbF_6^- (Figure 3A); with triflate (Tf^-) and BF_4^-, a polymerization is clearly observed but occurs much more slowly; with Cl^- and AcO^-, the polymerization is not observed, ii) the water content also enhances the polymerization (Figure 3A), iii) using palladium salts, a quite efficient process is found for $Pd(OAc)_2$ albeit the formulation remains liquid (Figure 3B); this can be ascribed to the formation of oligomers rather than a polymer i.e. in Figure 3D, the band characterizing the polyether formation at 1080 cm^{-1} is observed; however, this band is much weaker compared to that obtained for the DPSi/$AgSbF_6$ system; with $Pd(Cl)_2$, polymerization is not observed, iv) the comparison of palladium vs. silver salts initiating ability (Figure 3C) using a similar counter anion (AcO^-) shows that the Pd derivative is more efficient than Ag; unfortunately, a palladium salt with a very low nucleophilic anion (such as SbF_6^-) is not commercially available ruling out the development of very efficient system based on Pd salts.

Table 1. Characterization of the PCL formed in the presence of different DPSi/$AgSbF_6$/water initiating systems (conversion >96%). See text

entry	Ph_2SiH_2 (% w/w)	$AgSbF_6$ (% w/w)	Water (% w/w)	M_n (NMR)	M_n (SEC)[a]
1	1	1	-	9500	6770 (1.6[b])
2	0.7	0.7	-	13900	12040 (1.7[b])
3	1	1	0.5	3400	3470 (1.5[b])
4	0.7	0.7	0.7	2700	3360 (1.6[b])
5	1	1	0.05	6800	
6	1	1	0.1	5400	
7	1	1	0.4	3800	

[a] Polystyrene standards with a correction factor of 0.56. [17]

[b] Polydispersity indexes.

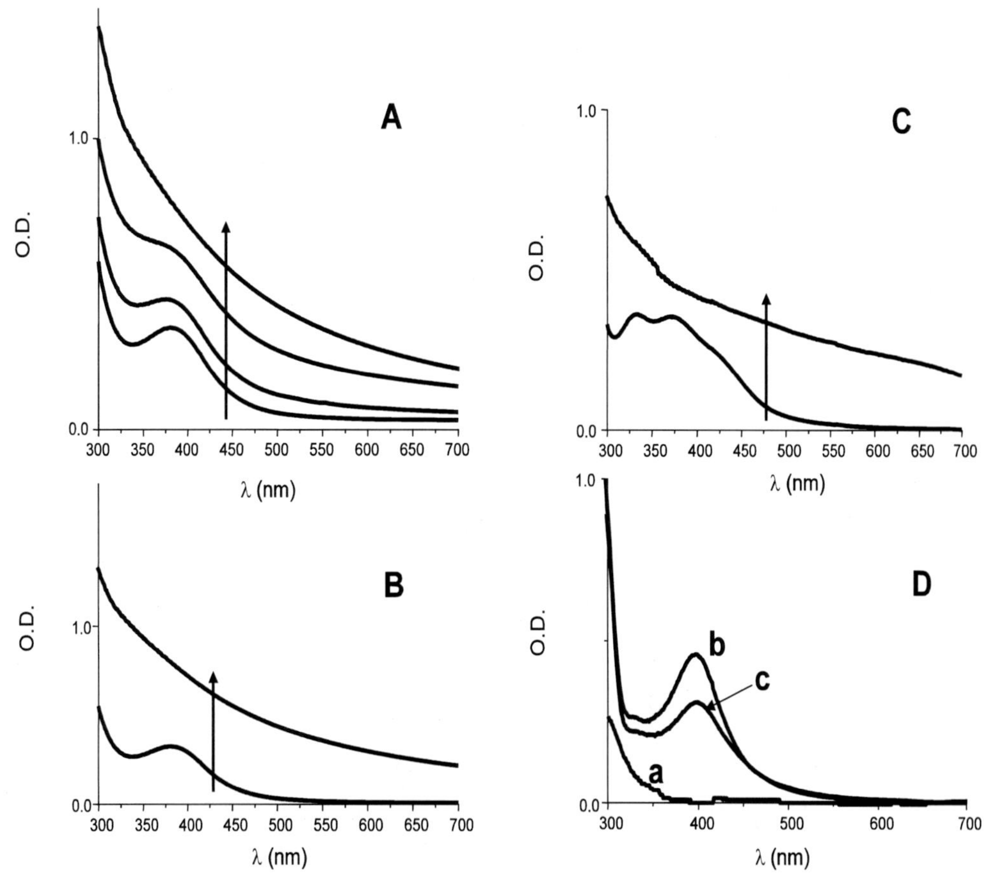

Figure 4. UV-visible absorption spectra: (A) $Pd(OAc)_2$ in acetonitrile upon addition of DPSi. (B) $Pd(OAc)_2$ in EPOX just before and after addition of DPSi. (C) $Pd(Cl)_2$ in EPOX just before and after addition of DPSi. (D) $AgBF_4$ in ε-caprolactone just before (a) and after addition of DPSi (b) followed by 36 hours of storage in the UV cell (c).

2. In-situ Incorporation of Nanoparticles

The in-situ formation of metal nanoparticles is demonstrated in solution by UV-visible spectroscopy (Figure 4). For PCL powders, this technique can hardly be applied. The addition of DPSi to $AgBF_4$ in ε-caprolactone ($AgBF_4$ was used instead of $AgSbF_6$ to avoid the formation of polymer and allow an easier characterization of Ag(0)) leads to the formation of Ag(0) nanoparticles well evidenced by UV-visible spectroscopy i.e. the surface plasmon resonance (SPR) band typical of the silver nanoparticles is observed at about 400 nm (Figure 4D). Unprotected Ag nanoparticles characterized by the size domain < 15 nm display a SPR band at a similar wavelength. [18] Remarkably, these particles are quite stable in ε-caprolactone i.e. after a 36 hours storage, only a modest decrease of the SPR band is observed. The absorption maximum remains also quite similar i.e. 398 vs. 402 nm at t = 0 and t = 36 hours (Figure 4D), respectively. This demonstrates that the particles size remains probably unaffected.

Mixing $Pd(OAc)_2$ or $Pd(Cl)_2$ with DPSi in acetonitrile reveals a gradual color change from yellow to brown. The initial absorption band at ~ 390 nm due to palladium acetate or palladium chloride is replaced by a monotonically broad spectrum typical of palladium nanoparticles albeit for Pd, this does not correspond to a SPR band. [19] This evidences the reduction of Pd(II) ions to Pd(0) (Figure 4A). A similar behavior is found when EPOX is used as a solvent demonstrating that Pd(0) nanoparticles formation also occurs (Figures 4B-C).

3. The initiation Process

Based on other redox reactions involving silanes, [14,20-22] processes (1) and (2) have already been proposed for the $R_3SiH/AgSbF_6$ interaction leading to a silylium cation alone or silylium cation and a Brönsted acid, (1) being assumed as the main reaction. On the other side, the formation of silylium cations when using chlorosilanes (R_3SiCl) instead of R_3SiH with silver salts (3) was well characterized. [23]

$$2\,R_3SiH + 2\,AgSbF_6 \rightarrow 2\,R_3Si^+SbF_6^- + 2Ag(0) + H_2 \qquad (1)$$

$$R_3SiH + 2\,AgSbF_6 \rightarrow R_3Si^+SbF_6^- + 2Ag(0) + H^+SbF_6^- \qquad (2)$$

$$R_3Si\text{-}Cl + AgSbF_6 \rightarrow\rightarrow\rightarrow R_3Si^+SbF_6^- + Ag(0) + \text{products} \qquad (3)$$

In our work, the role of reaction (3) is clearly outlined: $(TMS)_3Si\text{-}Cl/AgSbF_6$ (1%/1% w/w) initiating system leads to an efficient polymerization process (gel time about 4 hours at RT) in full agreement with the decisive role of $R_3Si^+SbF_6^-$ in the initiating mechanism. The formation of Ag(0) nanoparticles is also evidenced by through their SPR band observed at 401 nm. By analogy, we consider that reaction (1) corresponds to the major pathway of the R_3Si^+ production during the polymerization initiation process using $R_3Si\text{-}H/AgSbF_6$. With a Pd salt, such processes can also be expected with a double reduction of Pd(II) to Pd(0) as evidenced by the Pd(0) nanoparticles formation.

As silylium cations are very reactive species, their observation is very difficult. [23] Some radiochemical experiments were reported [24] but can hardly be applied for the chemical processes investigated here: therefore, an indirect approach based on the reaction of these species with water was used (4). Reaction (2) was neglected compared to (1) as evidenced in [20-22]. The addition of DPSi (0.071 M) to a solution of silver salt in water ($AgBF_4$ 0.072 M) leads to a strong decrease of the pH (from about 5 to ~ 2.15 i.e. $[H^+] = 0.07$ +/- 0.01 M). This equivalence between the initial silane and final acid concentrations demonstrates that the yield of R_3Si^+ formation from silane is close to 1.0.

$$R_3Si^+ + H_2O \rightarrow R_3SiOH + H^+ \quad (4)$$

In the absence of water, the initiating mechanism is related to silylium cations. the interaction energy between a silylium cation and ε-caprolactone was evaluated by density functional theory DFT calculations. A high reaction enthalpy is found (-47.5 kcal/mol) for reaction (5) demonstrating the high intrinsic affinity of R_3Si^+ for the carbonyl function of ε-caprolactone with the formation of a silylcarboxonium ion: this is also in agreement with the reported ability of such silylcarboxonium ions (generated by other approaches) to initiate the ROP of ε-CL. [6]

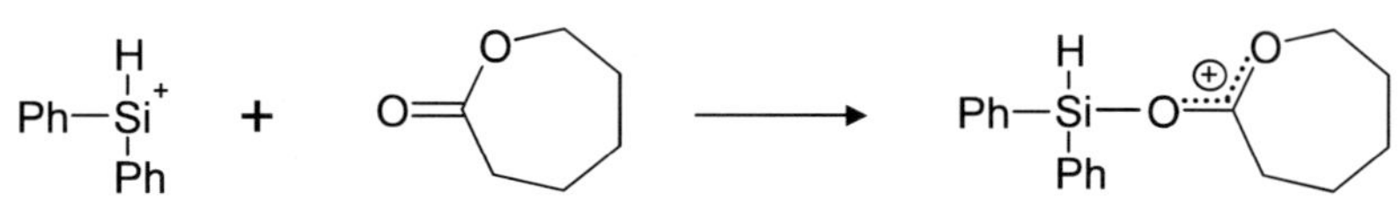

(5)

Upon addition of water (or presence of residual water in starting compounds), R_3Si^+ is converted into a Brönsted acid by (4) which can further participate to the initiation process (6), its contribution being dependent on the competitive reactions (4 vs. 7 where CL is used instead of ε-CL for sake of clarity). In these experiments, the reaction of silylium with H_2O is probably very fast (4) and the initiation mechanism is mainly associated with the acid formed. This is in full agreement with the observation of carboxylic chain end units, as revealed by matrix-assisted laser desorption/ionization (MALDI) mass spectrometry. The mass spectrum obtained after MALDI of the sample, using sodium iodide as the cationizing agent, displays a multi-modal distribution of singly charged PCL molecules, as indicated by both peak isotopic pattern and the 114 Da spacing between peaks (inset of Figure 5). From experiments performed with different salts (data not shown), signals observed in the mass spectrum of Figure 5 could be assigned to sodium adducts of PCL molecules, $[\alpha\text{-PCL}_n\text{-}\omega\text{+Na}]^+$. As a result, the sum of the end-group masses could be calculated from m/z values and was found to be $m_\alpha+m_\omega = 18$ Da, corresponding to end-groups such as –H and –OH. The presence of an acidic termination is fully consistent with the detection of a second minor PCL distribution, observed with a 22 Da mass shift as compared to the main one in Figure 5, which would correspond to sodium adducts of H-PCL_n-ONa. The presence of multiple envelopes within the same polymeric distribution is typical of a polydisperse sample, as further confirmed by distribution parameters calculated from these MS data ($M_n = 2415$ g mol^{-1}, $M_w = 3440$ $gmol^{-1}$, PDI = 1.4 in quite good agreement with SEC and NMR data – entry 4 in Table 1).

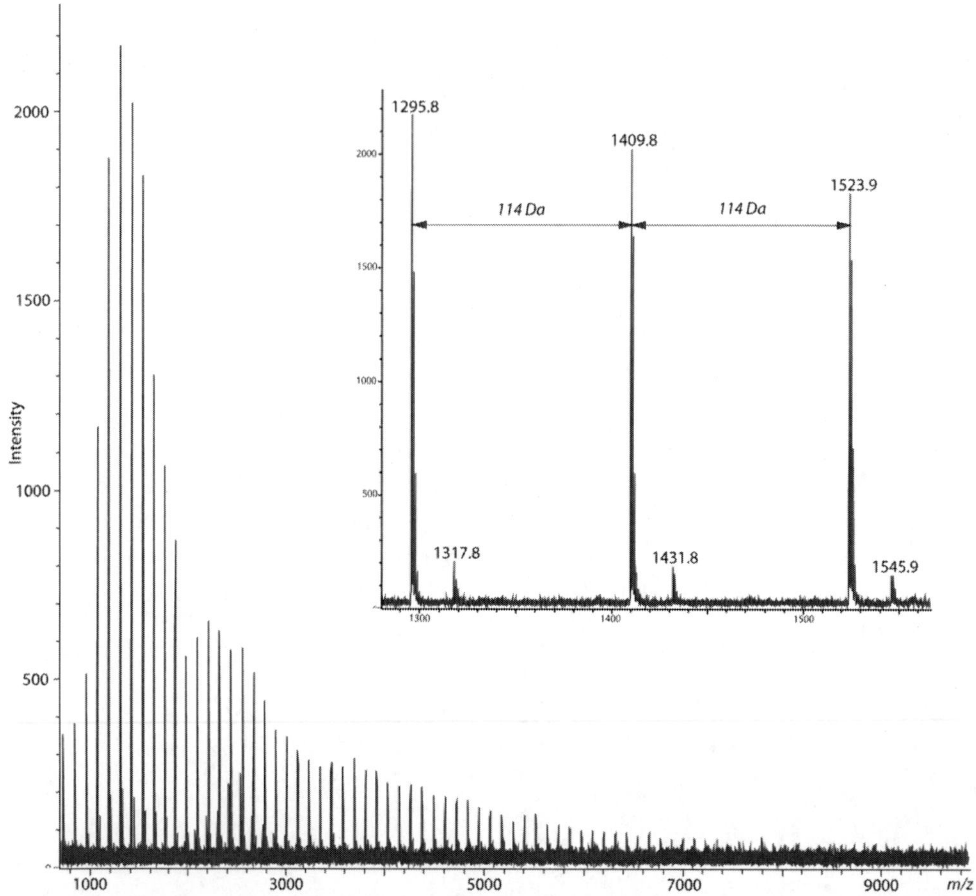

Figure 5. MALDI MS spectrum (reflectron mode, laser fluence : 40%) of the PCL sample, using DCTB as the matrix and NaI as the cationization agent. Inset : enlargement of the MALDI spectrum on the 1280-1570 *m/z* range.

The decrease of the gel time with [water] (which reflects an increase of the rate of polymerization) can also involve a chain transfer (8) as already observed for the cationic polymerization of epoxy groups: [25]

$$H^+ + CL \rightarrow H\text{-}CL^+ \quad (6)$$

$$R_3Si^+ + CL \rightarrow R_3Si\text{-}CL^+ \quad (7)$$

$$PCL^+ + H_2O \rightarrow PCL\text{-}OH + H^+ \quad (8)$$

The initiating role of alcohols when used in the presence of acids (as catalysts) in the PCL synthesis has been proposed in [2,7-8] as reactions (6, 9) where CL_{lin} stands for the open ε-CL unit.

$$H\text{-}CL^+ + ROH \rightarrow H^+ + R\text{-}O\text{-}CL_{lin}\text{-}H \quad (9)$$

(ROH can be also H_2O).

The polymerization proceeds here more slowly in the presence of isopropanol than upon addition of H_2O (see above in Figure 2A); as for the addition of water, the M_n linearly increases with the conversion. From an analysis of the ^1H peaks characterizing the polymer chain ends in the NMR spectra (ratio of the NMR peaks characterizing ^1H (a') at about 1.4 ppm vs. ^1H (b') at about 3.7 ppm; see scheme 3), it is calculated that about 60% of PCL chains are initiated by isopropanol according to (9). It can be assumed that the other PCL chains are initiated by the silylium cation (with or without the involvement of residual water; see above). Therefore, the effect observed here can indicate that i) traces of acid (as catalyst) are generated in our system (see reaction 4) and/or ii) the silylium cation can play a similar catalyst role as the Brönsted acid in (9).

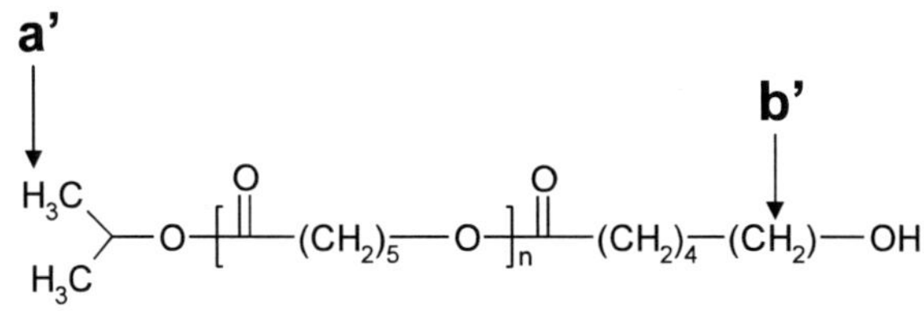

Scheme 3.

Conclusion

In the present study, a new initiating redox system based on a silane/Ag salt couple is described for the polymerization of ε-caprolactone according to a one pot procedure. High polymerization rates are noted under air and in simple glass vials. [26] According to a mimic situation in solution, silver nanoparticles should be present in the final PCL film. Remarkably, compared to other works where a heating is required, the reaction is carried out here at RT. The water content influences both the polymerization efficiency and the M_n and it allows an interesting tuning of the M_n. Using a Pd salt with a suitable anion should likely allow an efficient polymerization of ε-CL and an in-situ incorporation of Pd nanoparticles into the PCL film. The outstanding ability of these redox systems in the cationic thermal polymerization of other monomers and in cationic photopolymerization reactions will be presented in forthcoming studies.

References

[1] a) *Macromolecular engineering: from precise macromolecular synthesis to macroscopic materials properties and applications,* Matyjaszewski, K.; Gnanou, Y.; Leibler, L. Eds, Wiley-VCH, Weinheim, 2007. b) Crivello, J.V. *Photoinitiators for Free Radical, Cationic and Anionic Photopolymerization,* 2nd Ed.; Bradley, G., Ed.; New York, 1998. c) Crivello, J.V. Ring-Opening Polymerization, Brunelle, D.J. Ed.; Hanser: Munich, 1993. d) Fouassier, J.P. *Photoinitiation, Photopolymerization and Photocuring*: Fundamental and Applications; Hanser Publishers: New York, 1995. e)

Radiation Curing in Polymer Science and Technology; Fouassier, J.P.; Rabek, J.F. Eds. Elsevier Science Publishers Ltd.: London, 1993.

[2] a) Sando, F.; Sanado, H.; Shibasaki, Y.; Endo, T., *Macromolecules,* 2002, 35, 680-683. b) Lou, X.; Detrembleur, C.; Jérôme, R., *Macromolecules,* 2002, 35, 1190-1195. c) Persson, P.V.; Schroder, J.; Wickholm, K.; Hedenström, E.; Iversen, T., Macromolecules, 2004, 37, 5889-5893. d) Gazeau-Bureau, S.; Delcroix, D.; Martin-Vaca, B.; Bourissou, D.; Navarro, C.; Magnet, S., *Macromolecules,* 2008, 41, 3782-2784.

[3] Kamber, N.E.; Jeong, W.; Waymouth, R.M.; Pratt, R.C.; Lohmeijer, B.G.G.; Hedrick, J.L., *Chem. Rev.,* 2007, 107, 5813-5840.

[4] Hofman, A.; Szymanski, R.; Slomkowski, S.; Penczek, S. *Makromol.Chem.,* 1984, 185, 655-667.

[5] Kricheldorf, H.R.; Jonté, J.M.; Dunsing, R. *Makromol. Chem.,* 1986, 187, 771-785.

[6] Olah, A.; Wang, Q.; Li, X.; Rasul, G.; Prakash, G.K. *Macromolecules,* 1996, 29, 1857-1861.

[7] Shibasaki, Y.; Sanada, H.; Yokoi, M.; Sanda, F.; Endo, T. *Macromolecules,* 2000, 33, 4316-4320.

[8] Kunioka, M.; Wang, Y.; Onozawa, S-Y. *Polymer Journal,* 2003, 35, 422-429.

[9] a) Uygun, M.; Kahveci, M.U.; Odaci, D.; Timur, S.; Yagci, Y. Macromol. Chem. Phys., 2009, 210, 1867-1875. b) Dawson-Andoh, B.; Matuana, L. *European J. of Wood and Wood Products*, 2007, 65, 331. c) Yagci, Y.; Sangermano, M.; Rizza, G. Polymer, 2008, 49, 5195-5198. b) Sangermano, M.; Yagci, Y.; Rizza, G. *Macromolecules,* 2007, 40, 8827-8829.

[10] a) Crivello, J.V.; Lam, J.H.W. J. Polym. Sci. Part A.: Polym. Chem. 1981, 19, 539-548. b) Crivello, J.V.; Lee, J. *Makromol. Chem.* 1983, 184, 463-473.

[11] a) Onen, A.; Yagci, Y. Polymer, 1997, 38, 1423-1425. b) Onen, A.; Denizligil, S.; Yagci, Y. Angew. *Makromol. Chem.,* 1997, 245, 149-157.

[12] Crivello, J.V. J. *Polym. Sci. Part A.: Polym. Chem.,* 2009, 47, 5639-5651.

[13] a) Crivello, J.V. J. Polym. Sci. Part A.: Polym. Chem., 2009, 47, 1825-1835. b) Molleo, M.; Crivello, J.V. Macromolecules, 2009, 42, 3982-3991. c) Crivello, J.V. *Applied Org. Chem.,* 2010, 24, 169-178.

[14] Souane, R.; Tehfe, M.A.; Lalevée, J.; Gigmes, D.; Fouassier, J.P. Macromol. Chem. Phys. 2010, DOI: 10.1002/macp.201000002.

[15] Beuermann, S.; Paquet, D.A.; McMinn, J.H.; Hutchinson, R.A. *Macromolecules* 1996, 29, 4206-4215.

[16] a) Coleman, M.M.; Zarian, J. J. Polym. Sci. B, 1979, 17, 837-850. b) Elzein, T.; Nasser-Eddine, M.; Delaite, C.; Bistac, S.; Dumas, P. J. Colloid & Interf. Sci., 2004, 273, 381-387. c) Maier, M.A.R.; Gohy, J.F.; Fustin, C.A.; Schubert, U.S. *J. Am. Chem. Soc.* 2004, 126, 11517-11521.

[17] Save, M.; Schappacher, M.; Soum, A. *Macromol. Chem. Phys.,* 2002, 203, 889-899.

[18] a) Taleb, A.; Petit, C.; Pileni, M.P. J. Phys. Chem. B, 1998, 102, 2214-2220. b) Link, S.; El-Sayed, M.A. Annu. Rev. Phys. Chem. 2003, 54, 331-366. c) Pastoriza-Santos, I.; Liz-Marzán, L.M. Pure Appl. Chem., 2000, 72, 83-90. d) Van Hyning, D.L.; Klemperer, W.G.; Zukoski, C.F. Langmuir, 2001, 17, 3120-3127. e) He, S.; Yao, J.; Jiang, P.; Shi, D.; Zhang, H.; Xie, S.; Pang, S.; Gao, H. *Langmuir,* 2001, 17, 1571-1575.

[19] a) Scott, R.W.J.; Ye, H.; Henriquez, R.R.; Crooks, R.M. Chem. Mater. 2003, 15, 3873-3878. b) Creighton, J.A.; Eadon, D.G. *J. Chem. Soc. Faraday Trans.* 1991, 87, 3881-3891.

[20] Eaborn, C. *J. Chem. Soc.* 1955, 2517-2521.

[21] Marciniec, B. Z. *Anorg. Allg. Chem.* 1982, 495, 232-242.

[22] a) Ouellette, R.J.; Marks, D.L.; Miller, D.; Kesatie, D. J. *Org. Chem.,* 1969, 34, 1769-1771. b) Ouellette, R.J.; Marks, D.L. *J. Organomet. Chem.,* 1968, 11, 407-413.

[23] Lambert, J.B.; Kania, L.; Zhang, S. *Chem. Rev.,* 1995, 95, 1191-1201.

[24] Kochina, T.A.; Vrazhnov, D.V. *Glass Physics & Chem.* 2009, 35, 443-454.

[25] Hartwig, A.; Schneider, B.; Luhring, A. *Polymer,* 2002, 43, 4243-4250.

[26] In a recent ASAP paper (A. Parssinen, M. Kohlmayr, M. Leskela, T. *Repo, Polym. Chem.;* 2010), a highly efficient catalytic ROP of ε-CL under air (using titanium alkoxides) was achieved but it requires to work at 100-140 °C.

In: Silane: Chemistry, Applications and Performance
Editors: Katsuo Moriguchi and Susumu Utagawa
ISBN: 978-1-62257-432-2

Chapter 5

MODIFICATION OF NANOSILICA SURFACE BY METHYL METHACRYLATE SILANE COUPLING AGENTS IN A PSEUDO-LIQUID STATE

Igor Telegeev, Evgenij Voronin and Evgenij Pakhlov
Chuiko Institute of Surface Chemistry,
National Academy of Sciences of Ukraine, Kyiv, Ukraine

ABSTRACT

Composites based on silicas and methyl metacrylate silane coupling agents are advanced materials for resin modification, thermoplastics, glass fiber reinforced epoxy resins, sensor technologies. The purpose of the present research is to combine silica-organic compounds with inorganic constituents to tailor the binding properties. It deals with a new method of covering silica nanoparticles by methyl methacrylate silanes in a pseudo-liquid state. The adsorption process and the chemical reactions of metacrylate-containing silanes and the surface of nanosized silica were examined by means of infrared spectroscopy, thermal analysis and transmission electron microscopy (TEM). The FTIR analysis confirmed the successful introduction of functional groups onto the nanosilica surface. The ratio of the parent free silanol and the grafted organosilyl groups was determined to be 1.4:1, i.e. 55% of the silane molecules react with the surface silanol groups in a monofunctional and 45% in a bifunctional way.

Keywords: Nanosized silica; surface modification; methyl methacrylate silanes

INTRODUCTION

Highly dispersed nanosized silica (nanosilica) is used as a filler separately to create nanocomposites [1] and in the hybrid composite dental fillings [2]. To impart to nanosilica essential chemical properties such as polymerophilicity, adhesion, ability of matrix to cross-linking, etc., chemical modification of surface are used i.e. converting the surface silanol groups into organic or silica-organic ones by chemical reactions [3, 4].

Table 1. Some characteristics of nanosilicas

Nanosilicas	S_{BET}, m^2/g	d., nm	α_{OH}	
			mmol/g	*μmol/m*^{2}
A-300	320	8,5	0,65	2,0
A-175	170	16	0,38	2,2
A-100	109	25	0,24	2,3
A-70	68	38,5	0,19	2,8
OX-50	54	50	0,15	2,8

Table 2. Some parameters of the modifiers

Chemical name	Structure	T_b, ^0C	Molecular weight
Methyl methacrylate-methyldiethoxy silane (MMDES)	$(C_2H_5O)_2Si\text{-}(CH_2)\text{-}O\text{-}C\text{-}C\text{-}CH_3$ (Si–CH_3; C=O; C=CH_2)	79	232
Propyl methacrylate-trimethoxy silane (PMTMS) 3-Methacryloxypropyltrimethoxysilane (γ-MPTS)	$(CH_3O)_3Si\text{-}(CH_2)_3\text{-}O\text{-}C\text{-}C\text{-}CH_3$ (C=O; C=CH_2)	190	248
Propyl acrylate-tri-methoxy silane (PATMS)	$(CH_3O)_3Si\text{-}(CH_2)_3\text{-}O\text{-}C\text{-}CH$ (C=O; CH=CH_2)	68	234
Octyl-triethoxy silane (OTES)	$(C_2H_5O)_3Si\text{-}(CH_2)_7(CH_3)$	134	276
Methyl-triethoxy silane (MTES)	$(C_2H_5O)_3Si(CH_3)$	144	178

For the chemical modification of silica are most commonly used organosilicon compounds, and in these case hydrolytically stable surface groups are formed. Organosilicon compounds containing a hydrocarbon radical of active groups such as amine, epoxy, methacrylate, etc., tend to be volatile, so the modification is carried out from the liquid phase.

Liquid-phase method is versatile and allows you to modify almost any silica compounds. However, it has some significant drawbacks, mainly technological and environmental issues. First, is the necessity of using solvents which are often toxic or environmentally dangerous; secondly, is the removal of solvents due to the difficulties in the filtration process, and thirdly, is the presence of the stage of their disposal. Drying and grinding of the filtrate is heat-and energy-intensive processes that require additional costs.

A more progressive method for dispersed oxides is the gas phase modification method [10]. Gas-phase modification can be carried out under a reduced (in vacuum), with a normal (atmospheric) or at elevated pressure (autoclaving). The main requirement for the modifiers which also limits the wider use of gas-phase method is their volatility - which is not always the case.

Therefore, the aim of this study was to develop a method of nanosilica surface modification by low-volatility silane coupling agents in the absence of a liquid phase.

EXPERIMENTAL

In this work as fillers were used nanosilicas - Kalush plant (Ukraine) and the "Degussa" company (Germany) production some characteristics are summarized in Table 1.

The structures and names of silanes coupling agents chosen as modifiers are shown in Table 2.

The nanosilica modification was performed in a reactor of intensive mixing, which consists of a flask and a high-intensive blender. When rotated at 300-500 min^{-1}, the nanosilica transformed into a pseudo-liquid state. An amount of 10 g nanosilica was mixed for 10 min., and then a specific amount of the coupling agent was added to the mixture. The mixture was heated at different temperatures for a specific duration of time. After that, the heating was switched off, the sample left to cool, and then unloaded [11].

To study the functionalization of the silane coupling agent on the nanosilica surface, Fourier transform infrared (FTIR) spectroscopy was used. The degree of substitution of free silanol groups were determined from the ratio of optical densities of the absorption band of 3750 cm^{-1} before and after the reaction with silane (D_0 and D respectively) according to the formula:

$$\Theta = 1 - D/D_o,$$

where $D = \lg I_o/I$; a I_o and I – intensities of the transmitted infrared radiation ("baseline") and in the maximum of band 3750 cm^{-1}.

Spectral studies were performed on a spectrophotometer "Specord M80" (Carl Zeiss, Germany) and a Fourier transform infrared spectrometer, "Thermo-Nicolet" (Nicolet Instrument Corporation, USA). The studied silica samples were pressed into a plate of 8 × 28 mm weighing 20 ±0,5 mg in a special mold.

For the preparation of samples for transmission electron microscopy (TEM) analysis, a drop of nanosilica colloid solution was placed on a copper grid coated with a carbon membrane. The solvent then evaporated at room temperature leaving the nanosilica on the grid. The samples were examined with a TEM (SELMI PEM-125K, 100kV).

A thermal gravimetric analyzer (TGA) was used to measure the surface absorption of the nanosilica before and after surface treatment. The nanosilica powder was heated at room temperature to 1000 ^{0}C with a ramping rate of 20 ^{0}C/min. The weight loss of the nanosilica powder was recorded as a function of temperature.

RESULTS AND DISCUSSION

To determine the optimal interaction temperature of surface silanol groups with modifiers, the modification of nanosilica A-300 with γ-MPTS and MMDES silanes with stepwise increase in temperature and sampling of specimens for analysis at each stage of the process was carried out. In the reactor, at room temperature was injected 0.65 mmol/g of silane, based on the assumption that one silane molecule reacts with one surface silanol group. Then, the temperature in the reactor was increased stepwise to 50, 100 and 150 °C with exposure time at each temperature. The IR- spectra of the samples are shown in Figure 1.

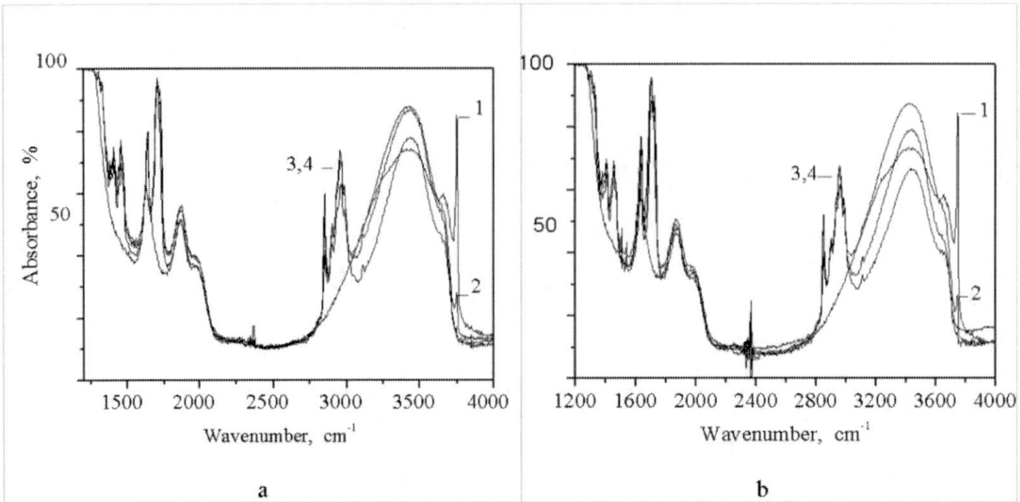

Figure 1. FTIR spectra of the nanosilica A-300 (α_{OH}= 0,65 mmol/g): initial (1) and after treatment with γ-MPTS (0,65 mmol/g) (a) and MMDES (0,65 mmol/g) (b) at 50 (2), 100 (3) and 150°C (4).

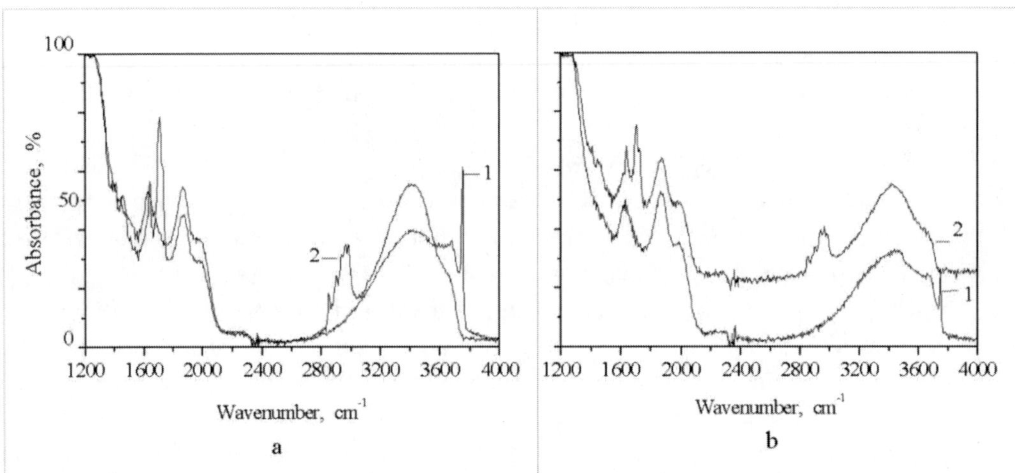

Figure 2. FTIR spectra of the nanosilica A-100 (a) and OX50 (b): initial (1) and after treatment with γ-MPTS at 100°C (2).

It is evident that even at 50 °C, almost 90% of the free silanol groups interact with the silane while at higher temperatures, reached their full replacement. Therefore, the modification at temperature 100 ° C can be considered as optimal.

At that temperature, a complete modification of all other nanosilicas by γ-MPTS silane was also achieved (Figure 2). This shows the complete availability of the surface of nanoparticles for chemical interaction with the molecules of the modifier, regardless of the degree of dispersion and aggregation of studied silicas. Note also, that for each sample, the input silane content of the reactor was equal to the surface free silanol groups (α_{OH}).

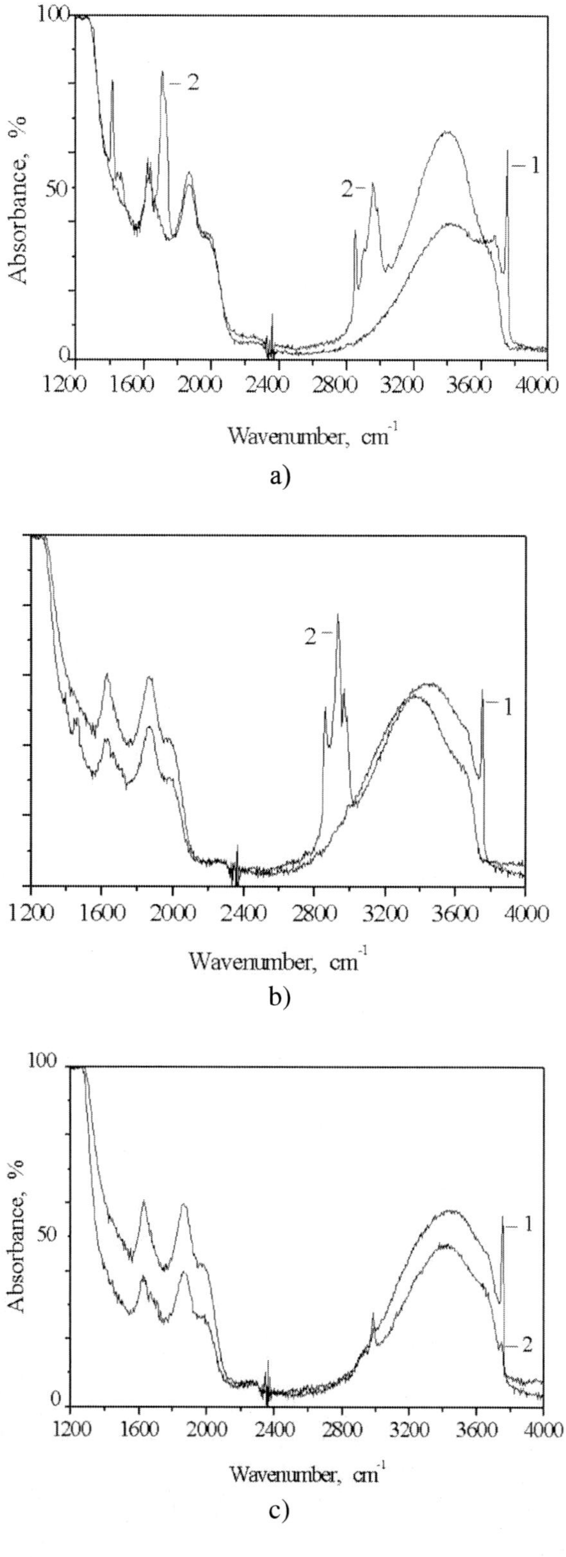

Figure 3. FTIR spectra of the nanosilica A-100: 1 – initial; 2 – and after treatment with PATMS (a), OTES (b), MTES (c).

According to literature data, alkyl alkoxy silanes containing various hydrocarbon radical electrondonor atoms (γ- aminopropyltriethoxysilane) typically interact with the surface of silica under milder conditions than alkyl alkoxy silanes [4]. This is explained by the fact that these atoms of one molecule of silane may be involved in the formation of an intermediate sorption complex of another molecule of silane, facilitating the transformation into the final reaction product, i.e., act as a so-called "Third component" - a catalyst for the modification [5].

For identification this trend observed under the conditions describing the modification to study the interaction of nanosilica A-100 with all the silanes was carried out using the same conditions - the same amount of modifier α_{OH} and the reaction temperature - 100 °. For all silanes except for the MTES, there is a complete replacement of silanol groups, except if the degree of substitution θ MTES was 83% (Figure 3).

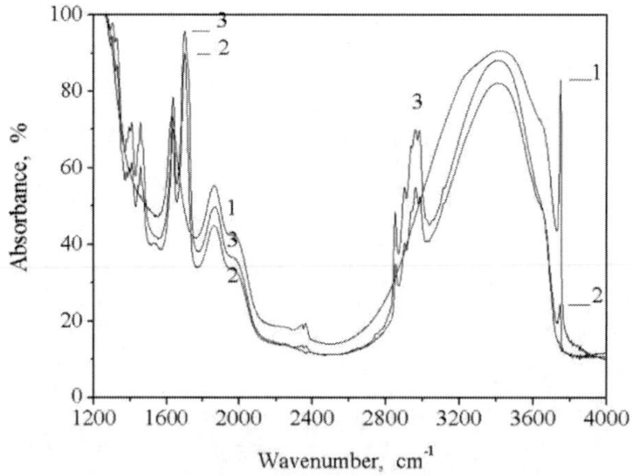

Figure 4. FTIR spectra of the nanosilica A-300 (α_{OH}= 0,65 mmol/g): initial (1) and after treatment with γ-MPTS at 100°C (0,4 mmol/g) (2) and (0,65 mmol/g) (3).

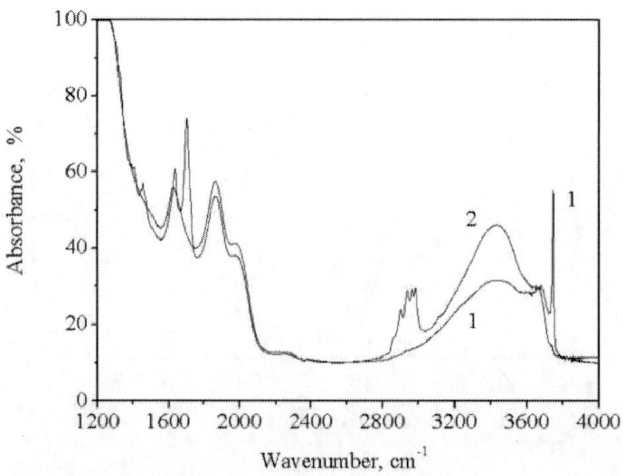

Figure 5. FTIR spectra of the initial nanosilica A-70 (α_{OH}= 0,19 mmol/g) (1) and after treatment with γ-MPTS (0,14 mmol/g) at 100°C (2).

We assume that the difference between the results obtained from published data is due to various conditions of modification. Data [4] were obtained in a completely or almost completely dehydrated vacuum samples. In our case, the silicas were the most hydrated state and the adsorbed water molecules played the role of "third component" [6] for all the modifiers, thereby, neutralizing the influence of electron-donating oxygen atoms of acrylate containing silanes. The lower activity of MTES is due to the kinetic factor and its greater volatility, thereby shifting the equilibrium surface - gas phase towards the latter.

We carried out the optimization of the modification temperature based on the ratio of silanol groups - silane molecule 1: 1. However, since the studied molecule of modifiers contain two or three alkoxy groups according to published data their interaction simultaneously with two or even with three silanol groups is possible [4, 7]. To clarify of this question, the modification of nanosilica A-300 by silane γ-MPTS was carried out, taken in amount of 0.4 mmol/g, which is about 62% of the α_{OH}. The substitution degree of silanol groups at the same time was 90% (Figure 4), i.e., an average of one molecule interacts with 1.45 OH γ-MPTS groups. Such parameter is possible in the case where 55% of the silane molecules react with surface-OH groups in a ratio of 1: 1 and 45% - 1: 2. Ratio of 1: 3, in our opinion, is unlikely [8]. This result agrees well with literature data. In [9], for the reactions of trimethylchlorosilane (TMHS) and $SiCl_4$ with the surface of dehydrated under-vacuum silica, reaction order 1.6 was obtained. This allowed the author to conclude that 40% of the molecules of silanes reacted in monofunctional and 60% - bifunctional-way.

In our opinion, the further evidence of the simultaneous implementation of mono- and bifunctional mechanisms of interaction of γ-MPTS with the silanol groups could be the results of the nanosilica A-70 modification.

In reactor containing the silica (α_{OH}=0.19 mmol/g) was added 0.14 mmol/g of γ-MPTS silane. If the above ratio of silane molecules reacting respectively in mono-and bifunctional scheme (55 and 45%) is true, so, with this amount of modifiers, we have to observe the complete replacement of the free surface silanol groups. Figure 5 shows that this degree of modification is almost achieved - the value of θ calculated from the infrared spectra was 0.96. The remaining 4% - is the experimental error.

The particle dispersion after treatment was evaluated by transmission electron microscopy (TEM). A reduced filler–filler interaction is manifested by the monodispersion of the silica filler as shown in (Figure 6).

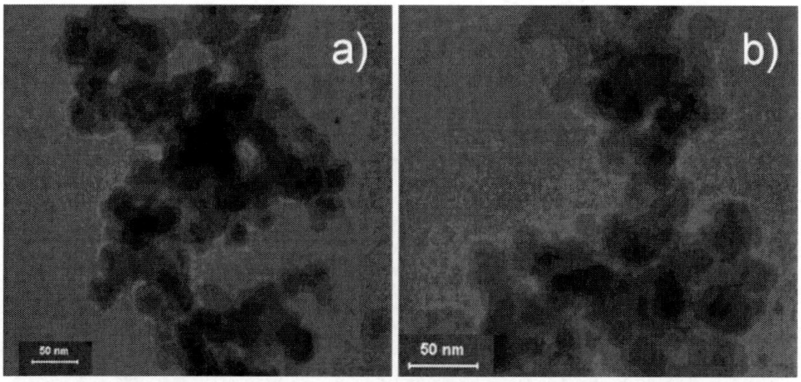

Figure 6. TEM pictures of nanosilica (a) and treated with γ-MPTS (b).

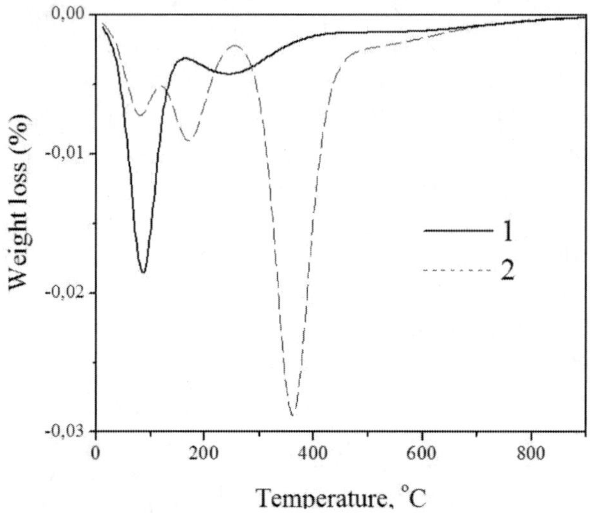

Figure 7. TG curves of the initial nanosilica (1) and after treatment with γ-MPTS (2).

The surface treatment of the nanosilica was also characterized by TGA. It can be seen that untreated nanosilica had a weight loss (about 2 wt %) below 200ºC, which is related to the elimination of physically absorbed water on the surface. The nanosized silica possesses a large surface area which makes it easy to pick up moisture in a typical laboratory environment even after they are dry. On the contrary, after silane surface treatment, the nanosilica evidently does not pick up water because of the changes in the hydrophilic behavior (Figure 7). The treated nanosilica steadily loses weight starting in the range of 300-500ºC, which is caused by the debonding and degradation of the grafted functional silane groups on the surface.

CONCLUSION

1. Designed and optimized way to modify nanosized silica by low-volatility silane coupling agents in a pseudo-liquid state.
2. It was shown that under experimental conditions, complete replacement of the free silanol groups by silane coupling agents for all nanosilicas regardless of their dispersion and aggregation was achieved.
3. The ratio of the parent free silanol and the grafted organosilyl groups was determined to be 1.4:1, i.e., 55% of the silane molecules reacted with the surface silanol groups in a monofunctional and 45% in a bifunctional way.

REFERENCES

[1] Basic Characteristics of Aerosil. Technical bulletin Pigments. Frankfurt: Degussa AG; 1997.

[2] Chuiko A.A. Silicas in medicine and biology. Kiev, Stavropol: Sat researcher proceedings; 1993.

[3] Kiselev. A.V. Intermolecular interactions in adsorption and chromatography. Moscow: High School; 1986.

[4] Tertyh V.A., Belyakova L.A. Chemical reactions involving silica surface. Kiev: *Science*. Dumka; 1991.

[5] Tertyh V.A, Pavlov V.V., Vatamanyuk V.I. On the influence of a third component to the reactions in the surface layer of silica. *Adsorption and adsorbents*, 1976, № 4, p.57-62.

[6] Tertyh V.A., Chuiko A.A., Mashchenko V.M., Pavlov V.V. Features of the silica surface chemisorption of trimethylchlorosilane. *J. Phys. Chemistry*, 1973, V.47, № 1, p. 158-163.

[7] Kiselev A.V., Lygin V.I. Infrared spectra of surface compounds. Moscow: *Nauka*; 1972.

[8] Tertyh V.A., Pavlov V.V. Problems of the reactivity of molecules attacking a fixed center. *Adsorption and adsorbents*, 1978, № 6, p.67-75.

[9] Hair M.L. Hydroxil groups on silica surface. *J. Non. Cryst. Solids*, 1975, V.19, p.299-309.

[10] Zou H., Wu S., Shen J. Polymer/silica nanocomposites: preparation, characterization, properties, and applications. *Chem. Rev.*, 2008, 108, p.3893-3957.

[11] Voronin E.F., Gun'ko V.M., Guzenko N.V., Pakhlov E.M., Nosach L.V., Leboda R., Skubiszewska-Zieba J., Malisheva M.L., Borysenko M.V., Chuiko A.A. Interaction of poly(ethylene oxide) with fumed silica. *Colloid Interface Sci.*, 2004, 279, p.326-340.

In: Silane: Chemistry, Applications and Performance
Editors: Katsuo Moriguchi and Susumu Utagawa
ISBN: 978-1-62257-432-2

Chapter 6

MODIFICATION OF ALGINATE MICROCAPSULES USING SILANE COUPLING AGENT AND APPLICATION TO ENZYME IMMOBILIZATION

Fumio Kurayama[1,*], Takeshi Furusawa[2], Masahide Sato[2], Newaz Mohammed Bahadur[3,4] and Noboru Suzuki[1,2,3]

[1]Center for Optical Research and Education, Utsunomiya University, Tochigi, Japan
[2]Department of Advanced Interdisciplinary Sciences, Graduate School of Engineering, Utsunomiya University, Tochigi, Japan
[3]Venture Business Laboratory, Utsunomiya University, Tochigi, Japan
[4]Department of Applied Chemistry and Chemical Engineering, Noakhali Science and Technology University, Noakhali, Bangladesh

ABSTRACT

This chapter presents a new and facile method for the modification of alginate microcapsules using 3-aminopropyltriethoxysilane (APTES) in aqueous media. Spherical-shaped microcapsules (Alg-capsule) with carboxymethyl cellulose liquid core and calcium alginate shell were prepared by a conventional method, and then, Alg-capsules were modified to produce APTES/alginate hybrid microcapsule (AP-capsule) without using any organic solvents or precipitating agents. The optical observation showed that the texture of AP-capsules was more glossy and transparent than that of Alg-capsules. The surface morphology and elemental compositions of microcapsules were characterized by Scanning Electron Microscopy (SEM), Fourier Transform Infrared Spectroscopy (FT-IR) and X-ray Photoelectron Spectroscopy (XPS). The results suggest that APTES molecules were incorporated to the framework of the alginate shells via electrostatic interaction between amino groups of APTES and carboxyl groups of alginate, and the hybrid shells had a thin and homogeneous structure. The shrinking of the capsule shells and the accumulation of APTES in the capsule shells were observed during the reaction between Alg-capsules and APTES molecules. This observation

*Corresponding author: Fumio Kurayama
Tel&Fax: +81-28-689-6163, E-mail address: fkurayama@cc.utsunomiya-u.ac.jp.

suggests the successful introduction of APTES molecules to the capsule shells as well as the self-condensation of loaded APTES molecules. In addition, the shell thickness and the amount of APTES in the capsule shells were found to depend on not only the pH of reaction solution but also the concentrations of APTES. From the permeability tests of compounds with different molecular weight (methyl viologen, nicotinamide adenine dinucleotide, diaphorase, bovine serum albumin), it was observed that the protein with 24 kDa easily permeated through Alg-capsule shells but hardly through AP-capsule shells. Therefore, AP-capsules were believed to have a molecular weight cutoff point of less than 24 kDa.

To demonstrate a practical application of AP-capsules, formate dehydrogenase (FDH) was selected as a model enzyme, and then the enzyme was immobilized with the capsules. The encapsulation of FDH in AP-capsules exhibited higher recycling ability than that in Alg-capsules due to the suppression of enzyme leakage. These results indicate that the AP-capsule is effective as a support of enzyme immobilization

INTRODUCTION

Microcapsules have recently attracted increased attention as microreactors that provide spatially restricted reaction fields for catalytic reactions. The microcapsules have some potential advantages such as protection of catalyst against the influences of outside environment and undesired reactants, promotion of cascade reactions in a limited space, controlled access of the reactants into the reaction area, easy separation and reuse. Various catalytic reactions such as synthesis of nanoparticle [1], photosynthesis [2], organic synthesis [3], biofuel synthesis [4, 5] and enzymatic synthesis [6] have also been successfully accomplished inside the microcapsules.

For preparing microcapsules, various kinds of polymers including polyelectrolytes [7], synthetic polymers [8], and biopolymers [9] have been utilized as a shell component since these polymers dominate characteristics capsule properties such as capsule size, shell thickness, permeability and rigidity. Among the polymers used for microencapsulation, alginate is one of the most commonly used as shell component due to their significant advantages such as good biocompatibility, low cost and simple preparation under mild conditions. The simplest method for preparing macroscopic alginate microcapsules (size in millimeters) is to use a single nozzle technique [10]. The alginate microcapsules obtained with a single nozzle can be briefly classified into two main types: (1) microcapsules with alginate liquid core and alginate gel shell (multiple-step process), and (2) microcapsules with non-gelling polymer core and alginate gel shell (one-step process). The former microcapsules are generally fabricated by dripping a sodium alginate solution into crosslinker solution such as calcium chloride ($CaCl_2$) solution and recovered prior to the complete gelation. In some cases, subsequent processes of coating additional protective layers on the surface and liquefaction process of alginate core are often recommended in order to improve the mechanical properties and stability against changes in the external environment.The preparation of the latter microcapsules consists in the dropwise addition of viscous solution containing crosslinker and non-gelling polymer (as a thickener) into the sodium alginate solution under high speed stirring, yielding spherical microcapsules with alginate shell and viscous liquid core [11]. The latter method is more flexible compared to the former one because several thickeners are available as a component of core solution, and it is possible to remove the thickeners such as carboxymethyl cellulose [12] and polyethylene glycol [13] from the microcapsules, thus reducing diffusion resistivity caused by the viscous core inside the microcapsules.

To date, considerable efforts have been devoted to the encapsulation of a variety of enzymes such as invertase [14], glucose oxidase [15], glucuronidase [16], glucose oxidase and catalase [17], and inulinase [18] for the aim of the construction of an immobilized enzyme reactor system. However, enzymes with molecular weight of 200 kDa or less leak out easily through the alginate shell during the preparation process [19] because of the large pore size of alginate matrices, which restricts the application range of alginate-based microcapsules.

In recent times, microcapsules with organic/inorganic hybrid shell have found increased application in immobilization of biocatalysts due to their good hydrophilicity, controllable transport characteristics, and good physicochemical stability. Effective methods for the modification of alginate-based microcapsules or beads with inorganic materials such as silica [20], hydroxyapatite [21], calcium phosphate [22], and calcium carbonate [23] have been extensively reviewed in the literatures. Among them, silicification approach mediated by biopolymer is the most attractive process, which proceeds at ambient temperature, pressure and pH values, producing a diversity of nanostructure frameworks. In general, there are two approaches for the formation of alginate microcapsules using silicon alkoxide and sodium silicate with organic/inorganic hybrid shell. One approach is based on the sol-gel process which involves hydrolysis and condensation of silicon alkoxide to form macromolecular siloxane networks. Sakai et al. [24] developed a facile method that involves interfacial sol-gel reactions of aminopropyltrimethoxysilane and tetramethyl orthosilicate at hexane/bead interface, leading to the formation of siliceous layer on the surface of alginate beads. The prepared beads with aminopropyl silicate/alginate composite shell membrane have been applied as a support for bioartificial pancreas [24], in vitro protein synthesis [25] and enzyme-entrapped magnetic beads [26]. Boninsegna et al. demonstrated that dripping of sodium alginate solution into $CaCl_2$ solution under a flow of silicon alkoxide induced the formation of siliceous layer at the gas/liquid interface, yielding calcium alginate beads enveloped by siliceous layer [27]. In these methods, it is necessary to use an organic solvent in order to obtain a homogeneous solution containing silicon alkoxides. Another approach is to utilize biological or synthetic polymers as templates and precipitating agents for biomimetic silicification. Poly-L-lysine [28, 29] or protamine [30, 31] was used as a precipitating agent of sodium silicate to inspire the formation of siliceous layer on the surface of calcium alginate beads or capsules. It was demonstrated that the obtained microcapsules were suitable for enzyme encapsulation and drug delivery application. The two different approaches described above opened new avenues to combine the superior physical properties of inorganic materials with alginate microcapsules. However, they need to use either organic solvent for dissolution of silicon alkoxide or precipitating agent for deposition of sodium silicate on capsule shells. From environmental and economic point of view, a process without the use of organic solvent and precipitating agent is preferable.

Aminofunctionalorganosilane, such as 3-aminopropyl triethoxysilane (APTES) is one of silicon alkoxides that has been extensively used as surface modifiers, coupling agents and additives [32-34]. Unlike most other organosilanes, aminofunctionalorganosilanes are known to have unique properties; aminosilanes are autocatalyzed for both hydrolysis and condensation owing to the base property of the amine functionality, and remain stable without precipitation in the form of oligomer in an aqueous solution [35-37]. In addition, aminosilanes have the potential to interact electrostatically with other materials having carboxyl and hydroxyl groups, and non-biocompatible alcohol released during hydrolysis of APTES can be eliminated before use.

Therefore, we have developed a new and facile method for preparing alginate/aminosilane hybrid capsule using 3-aminopropyltriethoxysilane (APTES) as an inorganic source [38]. In this method as illustrated in Figure 1, spherical-shaped capsules with carboxymethyl cellulose (CMC) liquid core and calcium alginate shell (Alg-capsule) were obtained in a conventional one-step gelling process, and fabrication of the APTES/alginate hybrid capsules (AP-capsule) was subsequently performed via the reaction between alginate and APTES molecules in an aqueous solution.

The morphological change of the microcapsules during the reaction between APTES and Alg-capsules as well as diffusional properties of the shell membrane have also been discussed in this chapter. Furthermore, enzyme immobilization was attempted to investigate the possible application of the AP-capsules.

The model enzyme introduced in the present study was formate dehydrogenase, which can catalyze oxidation of formate to carbon dioxide in the coupled reaction of oxidizedβ-nicotinamide adenine dinucleotide (NAD^+) to its reduced form (NADH) [39]. The activity and reusability of FDH encapsulated in AP-capsules was compared with that in Alg-capsules.

EXPERIMENTAL

Preparation of AP-Capsules

The preparation of AP-capsules was conducted using a home-made coaxial air-stripping extrusion device [38]. Briefly, 2 %(w/v) Sodium carboxymethyl cellulose (CMC, Sigma-Aldrich) solution containing 0.2 M calcium chloride ($CaCl_2$) and 2 U/ml formate dehydrogenase from Candida boidinii (FDH, EC1.2.1.2, Roche) as a core was prepared and dripped into a vigorously stirred 1%(w/v) sodium alginate (Kanto Chemical Co. Inc.) solution through the inner nozzle using a microsyringe pump.

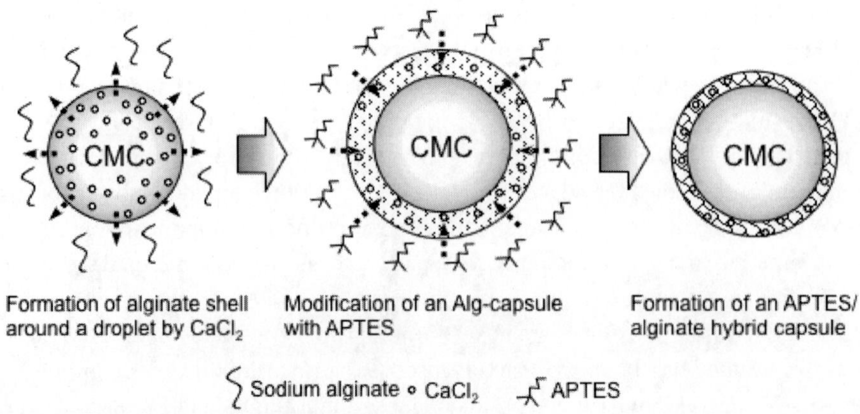

Figure 1. Schematic representation of the formation process of alginate capsule and the subsequent formation process of APTES/alginate hybrid capsule.

Nitrogen gas was passed through the outer nozzle in order to control the droplet size. After 2 minutes of dripping, the mixture was kept undisturbed for 1 min and then diluted three-fold by adding distilled water. After another 30 s, the alginate capsules were collected by filtration. The obtained alginate capsule is represented as Alg-capsule.

The formation reaction for APTES/alginate hybrid capsules was initiated at ambient temperature by putting the Alg-capsules into a 42 mM aqueous solution of APTES adjusted to pH 8.5 with diluted HCl. After 90 min, the capsules were filtrated and washed with water, and finally stored in water. The obtained APTES/alginate hybrid capsule is referred as AP-capsule.

In the case of crosslinking of AP-capsules by glutaraldehyde (GA), the prepared AP-capsules were incubated into 6ml of 0.05M MES (2-(N-morpholino)ethanesulfonic acid)) buffer solution (pH7) containing 1 %(v/v) GA for 60 min at room temperature, followed by washing with water.

The capsule size and shell thickness measurements were performed from capsule image captured by a digital camera, which was connected to a microscope. Outer and inner diameters of capsules were determined from the average diameters measured at two locations, and the measurements were performed on at least five capsules obtained under the same experimental conditions. The membrane thickness was calculated from the difference between the outer and inner diameters of the capsules.

To evaluate the number of mol of APTES incorporated into the shell of AP-capsules, the number of amino groups in an AP-capsule was determined using a colorimetric assay [38].

Diffusion Experiment

The microcapsules were added into a well-stirred 0.1 M MES buffer solution (3 ml, pH 7.0) containing one of the following substances: methyl viologen (MV^{2+}, Tokyo Chemical Industry Co., Ltd.), NAD^+ (Oriental Yeast Co., Ltd.), bovine serum albumin (BSA, Thermo Fisher Scientific Inc.) and diaphorase from Clostridium sp. (DAH, EC1.6.99., Toyobo Co., Ltd.). Changes in the concentration of each of the above substance were measured at 25 ºC as a function of time. All experiments were performed at least two times. The concentrations of BSA and DAH were determined by the Bradford method [40] using Coomassie plus protein assay reagent from Pierce. The concentrations of MV^{2+} and NAD^+ were measured from the absorbance at 258 nm and 260 nm, respectively, using spectrophotometer (Jasco, V-630).

Activity Measurement of FDH in AP-capsules

The catalytic activity of FDH encapsulated in AP-capsules was determined from the oxidation of formate coupling with the reduction of NAD^+ to NADH. An increase in the absorbance of NADH at 340 nm indicated the activity of FDH catalyst. Generally, NAD^+ and certain amount of FDH encapsulated AP-capsules were added into 8.85 mL of 0.1 M Tris-HCl buffer solution (pH 7.5). Then the mixture was kept undisturbed at 30 ºC for 10 min. After that,the FDH-catalyzed reaction was started by adding sodium formate into the reaction mixture. The concentrations of NAD^+ and sodium formate at the beginning of the reaction were 1 mM and 30 mM, respectively. The activity of FDH is defined as the initial rate of

NADH production by FDH. Based on the activity of FDH, enzyme immobilization yield and expressed activity were calculated. The immobilization yield is defined as the percentage of the activity of FDH in the core solution of AP-capsules with respect to that in the solution used in the preparation of AP-capsules. Expressed activity is defined as the percentage of the activity of the immobilized FDH with respect to the activity of the free FDH.

RESULTS AND DISCUSSION

Characterization of AP-capsules

Figure 2 shows pictures of Alg-capsules and AP-capsules. As can be seen, the texture of AP-capsule was found to be more glossy and transparent than that of Alg-capsules. Beari et al. reported that APTES in aqueous solution exists in the forms of intermediate entities such as silanetriol monomer, dimer, trimer and oligomer after the completion of hydrolysis and condensation of APTES molecules [34].

These hydrolyzed and condensed aminosilane molecules are stable over a period of up to one year. It is reported that alginate has the ability to interact with polycation having amino groups such as chitosan, poly-L-lysine and poly(allyl amine) via an electrostatic interaction [41-43]. The morphological changes between AP-capsules and Alg-capsules implied that during the reaction between Alg-capsules and APTES molecules, the soluble aminosilanes diffused toward Alg-capsules and adsorbed on alginate capsules via an electrostatic interaction between negatively charged carboxylate groups of alginate and positively charged protonated amino groups of APTES. The adsorbed APTES molecules were then hydrolyzed and condensed, leading to the formation of siloxane bridges with polymeric structure in the alginate matrix. Xu et al. also reported that alginate biopolymer could promote the hydrolysis and condensation of tetramethoxysilane to produce silica particles at ambient temperature and neutral pH [44].

It is also to be mentioned that there was no solid substances observed in the mixture during the reaction, suggesting that APTES only absorbed on the alginate matrix could undergo further hydrolysis and condensation reaction.

a b

Figure 2. Photographs of (a) Alg-capsules and (b) AP-capsules.

Figure 3 shows optical images of AP-capsules that were collected at different time intervals during the reaction between Alg-capsules and APTES. It is interesting to note that the outer diameter was reduced during the reaction whereas the inner diameter remained constant, indicating that only alginate shell was shrunk with time.This phenomenon did not occur in the absence of APTES in the reaction solution. From these observations, we assumed the aging of APTES molecules (condensation of near Si-OH groups), which resulted in the formation of SiO_2 network, leading to the capsule shrinkage. Therefore, there are at least two possible steps of the formation of APTES/alginate hybrid capsules, as shown in Figure 4: i) a hydrolyzed APTES molecule diffuses from a bulk solution to the alginate shell and adsorbs on alginate shell via an electrostatic interaction. ii) the hydrolyzed aminosilanes condensed to form a polysiloxane network in alginate shell, thus leading to the shrinking of the capsule shells.

More detailed structures of the hybrid capsules were observed using SEM. As shown in Figure 5(a), the surface of Alg-capsule shell exhibited a typical calcium alginate network structure [15]. In contrast, the surface of AP-capsule as shown in Figure 5(b) seemed to be smoother than that of Alg-capsule. The observation of the cross-section of AP-capsule shell as shown in Figure 5(c) revealed that the shell had a homogeneous structure and neither two-layered structure composed of alginate and siliceous layer [28, 45] nor a particle layer [27] was observed, suggesting that APTES was efficiently incorporated in the framework of alginate shells.

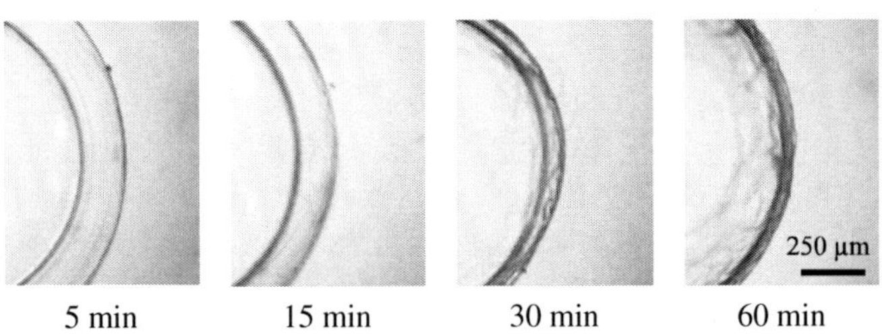

Figure 3. Time-course observation of an Alg-capsule during the formation reaction between Alg-capsule and APTES molecules.

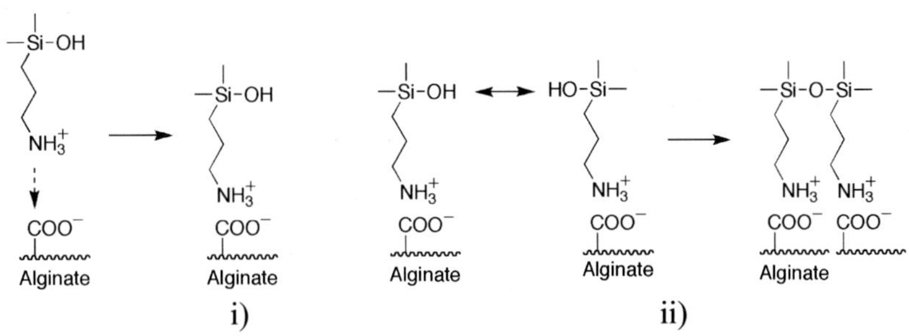

Figure 4. A possible mechanism of the formation reaction of APTES/alginate hybrid shells; (i) adsorption process via electrostatic interaction between protonated amino groups of APTES and carboxylate groups of alginate and (ii) condensation process to form polysiloxane network in alginate shells.

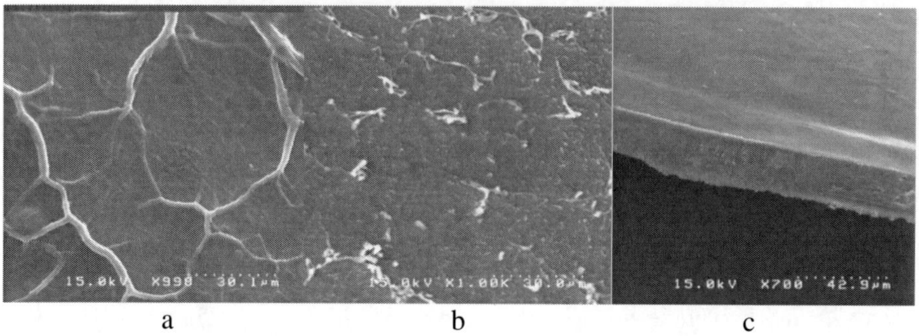

Figure 5. SEM images of capsules: (a) outer surface of Alg-capsule, (b) outer surface of AP-capsule and (c) cross-section of AP-capsule.

Chemical Composition of AP-capsules

The interaction between APTES and alginate shell was examined using FT-IR, which is reflected in Figure 6.In the FT-IR spectrum of sodium alginate (AlgNa), two strong absorption peaks at 1608 and 1417 cm^{-1} were assigned to the antisymmetric and symmetric stretching vibration of the carboxylate group, respectively. A strong and broad band observed at around 3424 cm^{-1} corresponds to O-H stretching. The weak band at 2930 cm^{-1} was assigned to aliphatic C-H stretching and two strong bands at 1096 and 1030 cm^{-1} were assigned to C-O stretching. The FT-IR spectrum of AP-capsule shell exhibited a strong and broad band centered at about 3430 cm^{-1} with a shoulder around 3350 cm^{-1}. This band was probably due to the overlapping of O-H and N-H stretching vibrations of alginate and aminopropyl groups, respectively. Compared to AlgNa, AP-capsules showed more intense bands at 2930 and 2860 cm^{-1} because the CH_2 stretching modes of the propyl chain in APTES was overlapped with the CH_2 bands of alginate shell in the case of AP-capsules [46]. In contrast to AlgNa, a new band appeared at 1517 cm^{-1} was attributed to the symmetric N-H bending vibration of protonated amino groups ($-NH_3^+$) [47]. In addition, a band in the range of 1250-1200 cm^{-1} in AP-capsules was observed due to Si-CH_2-R [48]. However, no characteristic band of Si-O-Si in the range 1130-1100 cm^{-1} was observed due to the overlay of Si-O-Si with the C-O stretching of alginate.

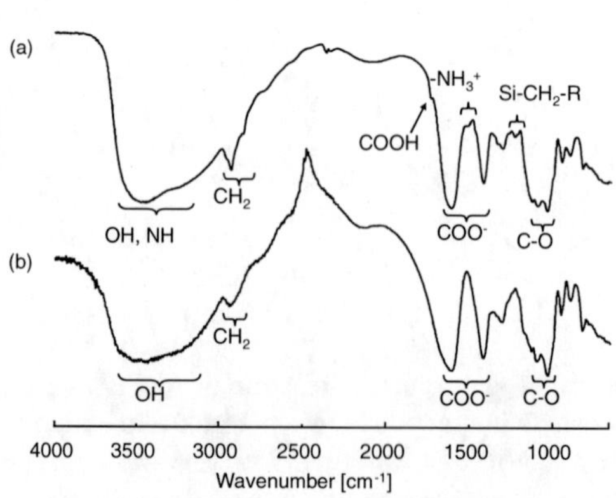

Figure 6. FT-IR spectra of (a) shell of AP-capsule and (b) sodium alginate.

In addition, a weak peak observed at 1735 cm^{-1} was attributed to the antisymmetric stretching mode of carboxylic acid [49]. Furthermore, no peak corresponds to the formation of covalent bonds such as amide bond (at 1486 cm^{-1})[50] was observed in the spectrum.

To further justify the incorporation of APTES in AP-capsule shell and determine chemical compositions of AP-capsule shell, inner and outer surfaces of an AP-capsule were analyzed by XPS. Figure 7 shows the survey scan spectra of the outer and inner surface of AP-capsules and Alg-capsules. Peaks correspond to elemental nitrogen (N) and silicon (Si) were observed at both surfaces of AP-capsule shells whereas these peaks was not observed in Alg-capsule shells.

The narrow XPS spectra of N1s for AP-capsule shown in Figure 8 indicate that the peak of the outer surface was almost identical in shape to that of the inner surface, and each peak was well fitted with two kinds of peak having a binding energy of 399.2 and 401.2 eV. The peak located in the region of 401.0-401.7 eV was assigned to the nitrogen of the protonated amine groups ($-NH_3^+$) of APTES, while the other peak located at 399.1±0.1 eV was attributed to the nitrogen of free amino groups ($-NH_2$) of APTES [48].

The atomic concentrations obtained from the XPS data is summarized in Table 1. As can be seen, the atomic concentration of calcium (Ca) in AP-capsule was significantly low compared to that in Alg-capsule although no difference in the atomic concentration of Ca was found between the outer and inner surface of AP-capsule and Alg-capsule.

Since the sum of the atomic concentrations of Ca and N (of NH_3^+) in AP-capsule was approximately equal to that of Ca in Alg-capsule, the difference in the atomic concentration of Ca between Alg-capsule and AP-capsule was probably due to the substitution of Ca^{2+}, which was complexed with carboxylate group of alginate by NH_3^+ of APTES. In addition, all elemental compositions for the outer and inner surfaces of AP-capsule shell were almost equal (Table 1). These results indicate that the formation of AP-capsule could be understood on the basis of electrostatic interaction between the protonated amino groups of APTES and carboxylate groups of alginate, and the APTES/alginate hybrid shell had a homogeneous structure.

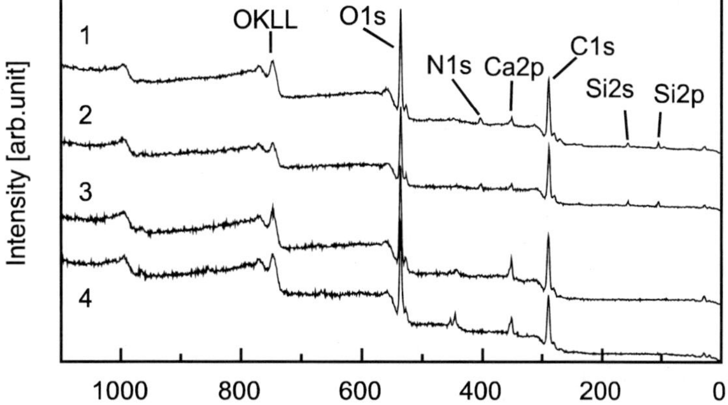

Figure 7. Survey XPS spectra of capsules shell; AP-capsule shell (1 outer surface and 2 inner surface); Alg-capsule shell (3 outer surface and 4 inner surface).

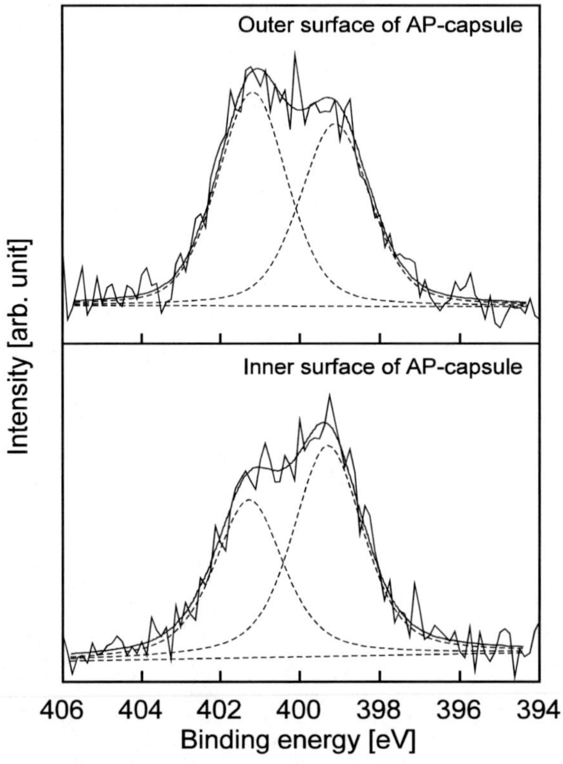

Figure 8. High-resolution XPS spectra of nitrogen (1s) of AP-capsule shell.

Table 1. Surface elemental compositions (in percentage) of outer and inner surfaces of AP-capsule and Alg-capsule

		C (%)	O (%)	Si (%)	Ca (%)	N (%)		NH_3^+ + Ca
						NH_3^+	NH_2	
AP-capsule	Outer surface	56.67	34.50	3.77	1.25	2.03	1.78	3.02
	Inner surface	60.40	30.62	3.93	1.20	1.66	2.19	2.78
Alg-capsule	Outer surface	59.24	37.65	0.00	3.11	0.00		–
	Inner surface	56.78	40.17	0.00	3.05	0.00		–

[a] Atomic concentrations of NH_3^+ and NH_2 were estimated from the curve fitted data of N1s peak as shown in Figure 8.

Formation of APTES/Alginate Hybrid Shell

The mechanism of the formation of AP-capsules was characterized from the extent of shrinking of alginate shell during the reaction. Figure 9 shows the effect of APTES concentrations on the shell shrinking of Alg-capsules during the formation of AP-capsules.

As can be seen, the shrinking of alginate shell progressed gradually with time and finally formed thin film. The shell-shrinking rate increased with increasing APTES concentration, suggesting that APTES aging process leads to shell-shrinking. According to Tanaka and Fillmore [51], the kinetics for swelling/shrinking behavior of gel can be described by the following equation:

$$\frac{d(t)-d(\infty)}{d(0)-d(\infty)} \approx \frac{6}{\pi^2}\exp\left(-\frac{t}{\tau}\right) \tag{1}$$

where d(t) is the gel diameter at time t and τ is the characteristic time for shrinking. Since all the curves obtained from the semi-logarithmic plot of the thickness versus time were fitted roughly to a straight line (data not shown), the time-dependent shrinking behavior can be assumed to follow Eq. (1). The dashed lines drawn in Figure 9 were calculated by using the characteristic times of shrinking obtained, and the results showed that shrinking behavior obeyed Eq.1.

In order to get more information about the formation mechanism of AP-capsule, the quantification of amino groups in AP-capsule was performed. The number of amino groups was then used to estimate the amount of APTES incorporated in the shell of Alg-capsule. Figure 10 shows the effect of initial concentration of APTES on the accumulation of APTES in Alg-capsules during the formation reaction. The rate of accumulation of APTES in Alg-capsules shell increased with increasing the initial concentration of APTES and reached a plateau for each condition. The saturated amount of APTES also increased with increasing the initial concentration of APTES. To estimate the kinetics of the reaction, the experimental data were approximated by fitting with first-order expression, and the rate constant was determined as 0.123 min^{-1}. The fitted curves are drawn using dashed lines in Figure 10 and it is shown that the accumulation of APTES in a capsule shell follows first order kinetics.

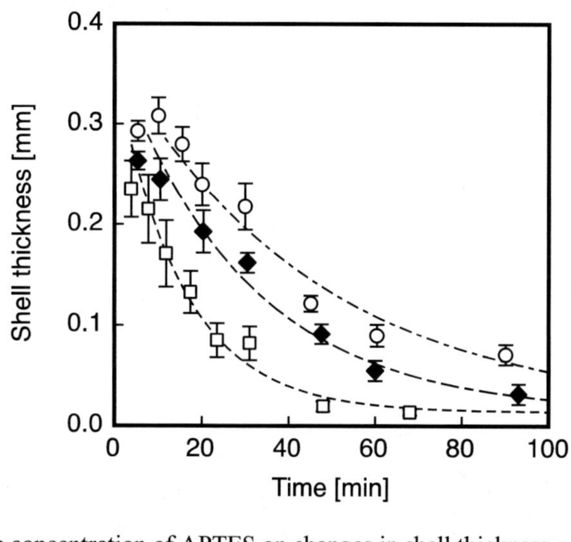

Figure 9. Effects of the concentration of APTES on changes in shell thickness with time during the formation reaction between APTES and alginate shells. APTES concentrations: 28 mM (O), 42 mM (◆), 70 mM (□). Error bars denote standard deviation.

Furthermore, it can be seen from Figure 9 and 10 that the shrinking of the capsule shell was in progress even after the accumulation of APTES in the capsule shell was completed. This fact also suggests that the aging process of the SiO_2 network continued even after the adsorption of the aminosilanes was completed, and led to the capsule shrinking.

Therefore, it can be concluded that higher accumulation of APTES molecules in alginate shells with time (as shown in Figure 10) corresponds to the stronger electrostatic interaction as well as faster condensation (as shown in Figure 9), which resulted in the shrinkage of alginate shell. These results demonstrated that the proposed reaction occurred only in the alginate shells and proceeded successfully in aqueous media without addition of either precipitating agents such as poly-L-lysine and protamine or any catalysts such as acid, alkali or fluoride salt.

Effect of pH on the Formation of APTES/Alginate Hybrid Shell

On the basis of the above mechanism, it can be considered that pH of the reaction mixture is an important factor for affecting the formation process of AP-capsules because the dissociation of carboxyl groups, protonation of amino groups and a sol-gel process involving hydrolysis and condensation of alkoxysilanes are well known to be influenced by the pH of the reaction solution. The effect of pH on the morphology of AP-capsules was examined by varying pH of APTES solution. Figure 11 shows photographs of AP-capsules obtained at 60 min of reaction time under different pH conditions. It was found that AP-capsules obtained at pH 6 and 9 had a thinner shell membrane than that obtained at the other pH conditions. The shell thickness of AP-capsules obtained at pH 10 was larger than that of Alg-capsules, suggesting that the amount of APTES adsorbed in the alginate shells was too small to induce the shell shrinking at pH 10. At around neutral pH region ranging from 7 to 8, the shell thickness of AP-capsules clearly changed depending on pH of the solution.

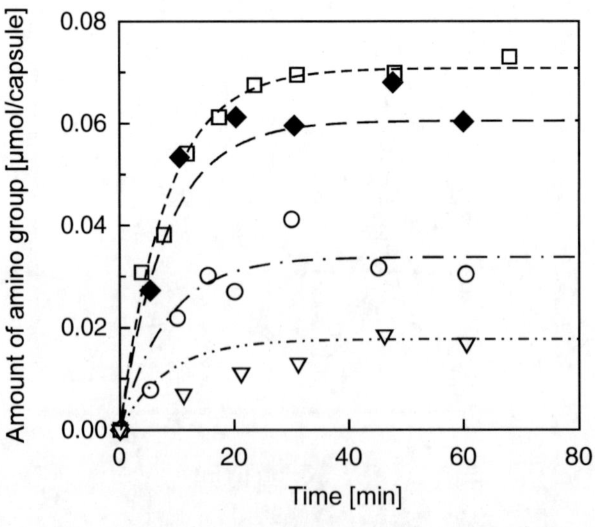

Figure 10. Effects of the initial concentration of APTES on the accumulation of APTES in alginate shell during the formation reaction between APTES and alginate shells. APTES concentrations: 14mM (▽); 28 mM (○); 42 mM (◆); 70 mM (□).

To clarify the pH-dependence of the shell thickness, the shrinking behavior of the capsule shells and the accumulation of APTES in the capsule shells were examined. Figure 12 shows shrinking rate (1/τ) of AP-capsule shell as a function of solution pH. As can be seen, the maximum rate of the shell shrinking was observed in the alkaline region (pH>8) while the shrinking rate decreased in the neutral pH region. Below pH 7, the shrinking rates were found to increase rapidly with decreasing solution pH. In contrast, no shrinking of alginate shells was observed above pH 11.

The similar tendency was also observed in the pH-dependence of both the adsorption rates and the saturated amounts of APTES in AP-capsule as shown in Figure 13. The carboxylate groups of alginate shells remained stable in the experimental pH range because the carboxylic acid dissociation constants (pKa) of mannuronic acid and guluronic acid in alginate are 3.38 and 3.56, respectively [52].

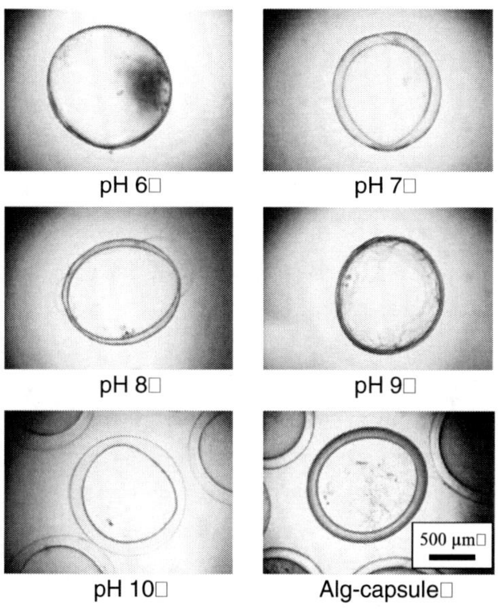

Figure 11. Photographs of AP-capsules obtained at different pH of the solution.

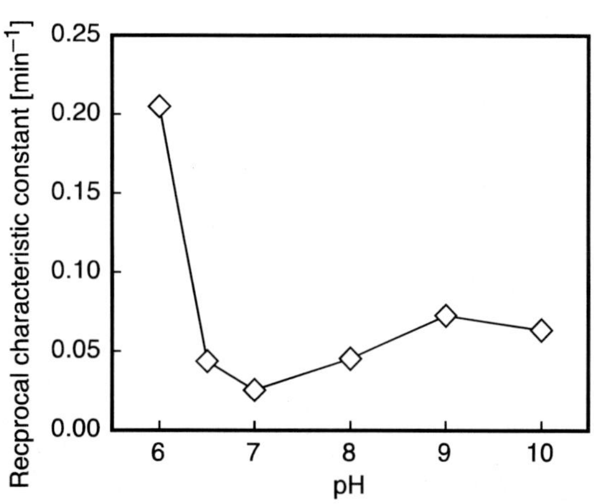

Figure 12. Effect of pH on the shell shrinking of capsules during the formation reaction between alginate shells and APTES.

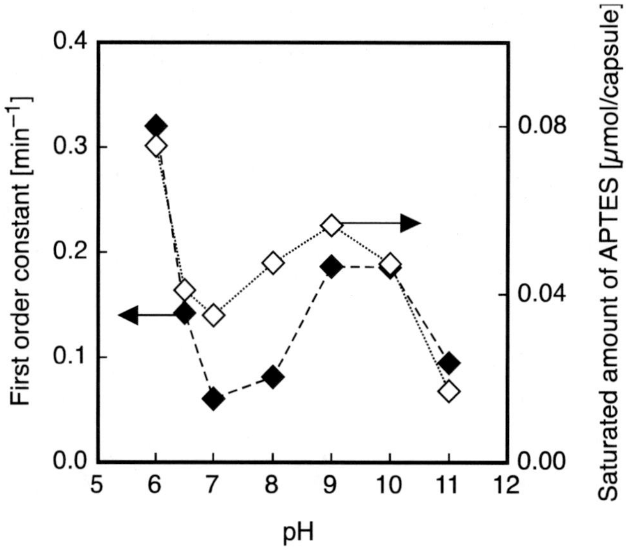

Figure 13. Effect of pH on the accumulation rates and saturated amounts of APTES in AP-capsules during the formation reaction between alginate shells and APTES.

Therefore, this pH-dependence of the shell thickness of AP-capsules could be explained on the basis of the dissociation of amino groups of APTES in the aqueous solution, which is as follows; since pKa of amino groups and silanol groups in APTES are known to be 9.6 and 6.7 [53], respectively, three different dissociative species of APTES molecules might exist in the solution depending on the pH of the solution. At above pH 10, the amino groups of APTES molecule were in their unprotonated form ($-NH_2$) whereas the silanol groups were dissociated substantially. As a consequence, APTES would hardly adsorb on the alginate shell matrix. At below pH 9, the amine groups of APTES were mostly protonated ($-NH_3^+$). However, APTES in water is known to form a zwitterion structure, and partial proton transfer from silanols to amino groups occurs, leading to the formation of intramolecular bond with a ring structure [35, 53]. Therefore, it was anticipated that APTES was partly in a zwitterion form with a ring structure at the pH condition between 7 and 8. In this pH range, the formation of zwitterions was expected to hinder the electrostatic interaction between APTES molecules and alginate shells. Furthermore, since the protonation of the zwitterions (protonation of silanolate groups) is occurred below pH 6.7 [53], the protonated amino groups in APTES would be increased, thus facilitating the incorporation of APTES into the alginate shell matrix. Since the shrinking rate of the capsule shell is associated with the amount of APTES incorporated in the capsule shell as discussed in Figures9 and 10, both the shrinking and accumulation rates shown in Figures 12 and 13 exhibited a same tendency for pH changes in the solution. These results revealed that the formation of AP-capsules was greatly influenced by pH of the reaction solution and promoted under acidic and alkaline conditions, and these behaviors also support that APTES/alginate hybrid shell was formed via the electrostatic interaction between protonated amino groups of APTES and carboxylate groups of alginate.

Effect of Molecular Sizes of Different Substrates on Their Diffusion Behavior to AP-capsules

For enzyme immobilization, microcapsules needs to have a selective permeable membrane that allows free passage of reactants and products while retaining the enzymes. Since the transport properties of microcapsules usually depend on a size exclusion or electrostatic repulsion of the shell membrane, diffusion behavior of solutes with different molecular sizes across capsule shells is often determined in order to characterize the transport properties such as diffusion coefficient of solutes, molecular mass cutoff and the solute ingress or egress percentages [54-57]. In this section, a study on the diffusion behavior of solutes with different molecular sizes has been demonstrated to characterize the transport properties of AP-capsules.

For examining diffusion behavior of large macromolecules, DAH with a molecular weight (MW) of 24 kDa [58] and BSA with a MW of 67 kDa were used as a model protein. Figure 14 shows diffusions of DAH and BSA into AP-capsules and Alg-capsules.

In the case of Alg-capsules, it was found that the concentration of DAH in solution gradually decreased with time and finally reached at equilibrium. In contrary, it can be seen that both the concentrations of BSA and DAH in bulk solution remained constant in the case of AP-capsule.The smooth and thin shell membrane of AP-capsule prevented such proteins from passing through the shell. From these results, the AP-capsules were found to have the ability to prevent proteins with a lower molecular weight of 24 kDa from passing through the capsule shells.

For examining the diffusion behavior of lower molecular weight substrates, MV^{2+} (MW of 257 Da) and NAD^+ (MW of 663 Da) were used as typical examples of enzyme substrates in diffusion experiments. MV^{2+} has usually been used as a redox mediator for electrochemical investigations of biological redox systems [59]. NAD^+ is a well-known coenzyme and used with many dehydrogenases and oxidoreductases [60]. Figure 15 shows the diffusion of MV^{2+} and NAD^+ into AP-capsules and Alg-capsules. As shown in Figure15, both NAD^+ and MV^{2+} gradually diffused from bulk solution to the capsules.

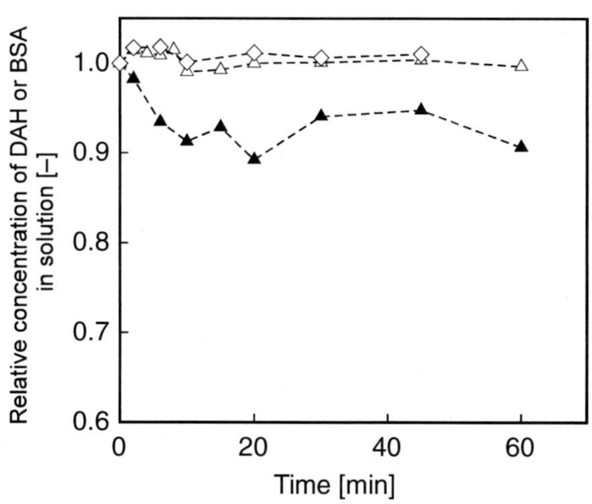

Figure 14. Diffusion profiles of substrates from the solution to Alg-capsules (DAH (▲)) and AP-capsules (DAH (Δ), BSA (◇)).

Compared to the Figures 14 and 15, it can be concluded that AP-capsule had a semipermeable membrane through which the protein with MW of 24 kDa could not diffuse whereas small substrates with low molecular weight (MV^+ and NAD^+) passed through the capsule shells.

Therefore, it is suggested that the use of aminosilane in complexation with alginate microcapsule is a promising strategy for enzyme immobilization.

One approach to modulate the transport properties of microcapsules is to use crosslinking agent such as glutaraldehyde (GA) [61]. GA is extensively used in protein immobilization, and aldehyde groups of GA react with amino groups, leading to the formation of Schiff base linkage [62].

Some studies have shown that treatment of microcapsules having amino groups with GA results in an improved stability and lower molecular cutoff point [63-65]. Therefore, the crosslinking of amino groups of AP-capsule with GA appears to be an attractive strategy for modulating the diffusion characteristics and molecular cutoff point. To verify the above assumption, AP-capsules crosslinked with GA (GA-capsule) were prepared and diffusion behavior of NAD^+ and MV^{2+} into GA-capsules was investigated. As shown in Figure16, the concentration of MV^{2+} in solution was gradually decreased and reached at equilibrium whereas the concentration of NAD^+ in a solution remained unchanged. Thus, we believe that the molecular cutoff point of GA-capsule exists between MW of 256 and 663.

The interesting feature of this semipermeable membrane of GA-capsule has suggested that NAD^+ might be encapsulated inside of the GA-capsules. Since NAD^+ is relatively expensive, various methods have been proposed to immobilize NAD^+ molecules or polymerized NAD^+ on various supports [60].

Our results suggested that GA-capsules might be a new support for the encapsulation of NAD^+ molecules. Therefore, the transport properties of AP-capsules could be further modified by crosslinking with GA.

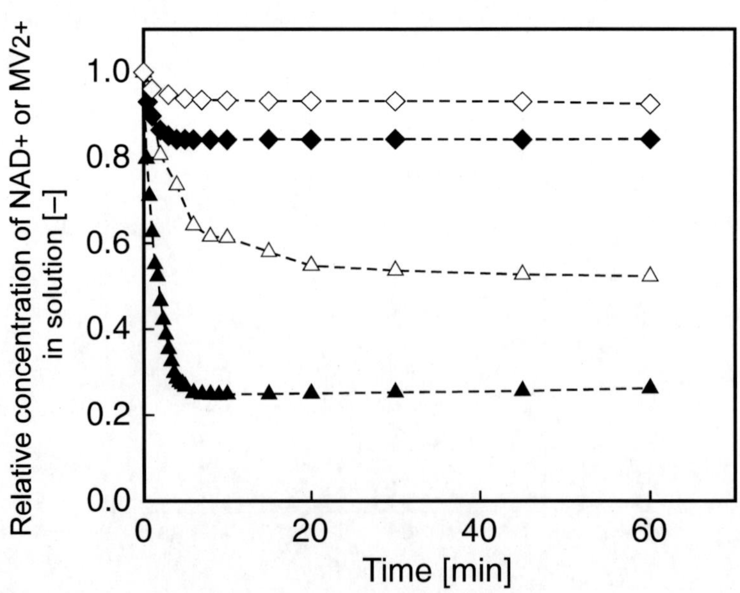

Figure 15. Diffusion profiles of substrates from the solution to Alg-capsules (MV^{2+} (▲), NAD^+ (Δ)) and AP-capsules (MV^{2+} (■), BSA (◇)).

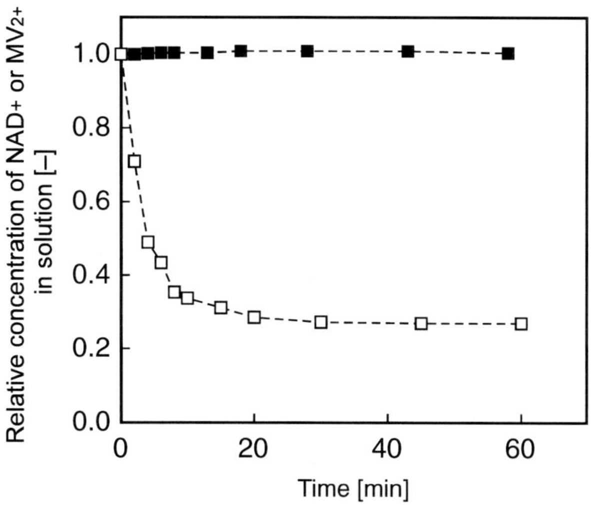

Figure 16. Diffusion profiles of MV^{2+} (□) and NAD^+ (■) from the solution to GA-capsules.

Immobilization of FDH in AP-capsules

To examine the feasibility of immobilizing enzyme in AP-capsules, FDH-loaded AP-capsules were applied in the enzymatic conversion of NAD^+ to NADH with a coupled oxidation of formate to CO_2.

Figure 17 shows time course of NADH production by FDH inside AP-capsules with different sizes in diameter. It can be seen that the amount of NADH produced per unit weight of the immobilized FDH was increased with time and decrease of microcapsule size. This result indicates that FDH was active inside AP-capsules.

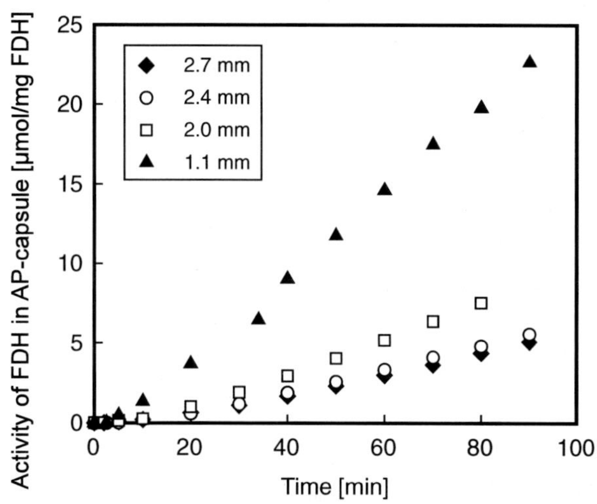

Figure 17. Changes of the activity of FDH in AP-capsules with different diameters.

Table 2. The effect of capsule diameter on expressed activity and immobilization yield

Diameter (mm)	Expressed activity (%)	Immobilization yield (%)
2.7	12	46
2.4	13	42
2.0	20	34
1.1	47	14

The results of immobilization and expressed activities were summarized in Table 2. It can be seen that with decreasing the microcapsule size, immobilization yields were decreased whereas expressed activities were increased.

Considering the results obtained in the diffusion experiments as mentioned above, we assume that the leakage of FDH might be mainly occurred during the preparation of Alg-capsules, in which FDH leaked out from droplets to sodium alginate solution before forming alginate shells around the droplets. Since a decrease of microcapsule size corresponds to an increase of surface area of a microcapsule, the higher amount of leakage of FDH from the microcapsules was probably due to the increase of surface area of a droplet containing FDH in the preparation process of Alg-capsules. On the contrary, it was found that decrease of the microcapsule size resulted in the increase of the expressed activity of FDH. Generally, enzymatic reactions inside microcapsules are limited by the diffusion of substrates. In this case, the consumption of substrates by the enzyme inside microcapsules led to the concentration gradient of substrates along radial direction. As a consequence, concentration of substrates was too low in the central part of a microcapsule that enzymes located at the center of the microcapsules could not work efficiently. Therefore, the apparent activity of enzyme in the microcapsules was lower than that in a solution. Based on this consideration, the decrease of capsule size corresponds to shortening diffusion distance of substrates, leading to the improvement of the expressed activity.

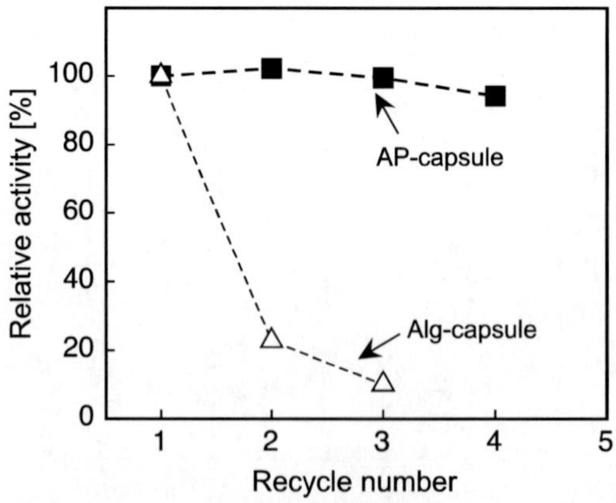

Figure 18. Recycling stability of FDH encapsulated in Alg-capsules and AP-capsules.

One important purpose of immobilizing enzyme is to facilitate catalyst recycling and reuse to reduce cost. The reusability of AP-capsules was demonstrated by examining loss in activity of AP-capsules during reusing cycles. Figure 18 illustrates the recycling stability of AP-capsules and Alg-capsules.

After four repeated batches, almost no loss in catalytic activity was found for AP-capsules whereas the rapid decrease in the catalytic activity was observed in the case of Alg-capsules. Since FDH is known to have a molecular weight of 74 kDa [66], the decrease of FDH activity in Alg-capsules was due to the leakage of FDH during the soaking, separating, and washing processes employed during recycling. The results showed that AP-capsule shell played a very important role in protecting the FDH from leaching and, the reusability of AP-capsules was increased significantly compared to that of Alg-capsules. Therefore, the AP-capsule is effective as a support of enzyme immobilization.

Conclusion

We have described a facile and effective approach for preparing the microcapsules with APTES/alginate hybrid shells (AP-capsule) based on the combination of a one-step gelling process with a sol-gel process. The AP-capsules were successfully produced in aqueous media without addition of either precipitating agents such as poly-L-lysine and protamine or any catalysts such as acid, alkali or fluoride salt.

The reaction between Alg-capsules and APTES molecules was found to induce shell shrinking and accumulation of APTES in the capsule shells, which depended on the concentrations of APTES as well as pH of the reaction solution. The permeability tests of different molecular weight compounds revealed that AP-capsules contain a semipermeable membrane that was permeable to small substrates such as NAD^+ and DAH but impermeable to proteins such as BSA and DAH. Moreover, it was found that the amino groups in AP-capsules were available for modulating the permeability of the membrane by crosslinking with GA. AP-capsules were found to be capable of encapsulating FDH, and indicated a significant recycling stability as compared to that in Alg-capsules. Therefore, it is anticipated that the AP-capsule could allow effective support for enzyme immobilization.

Acknowledgment

Special thanks to the Journal of Colloid and Interface Science for publishing a part of this research work.

References

[1] Shchulkin, D. G., Ustinovich, E. A., Sukhorukov, G. B., Möhwald, H., Sviridov, D.V. *Adv. Mater.* 2005, 17, 468-472.

[2] Shchukin, D. G., Sviridov, D. V. *J. Photochem. Photobiol.* C 2006, 7, 23-39.

[3] He, J., Yoneyama, Y., Xu, B., Nishiyama, N., Tsubaki, N. *Langmuir* 2005, 21, 1699-1702.

[4] Ylitervo, P., Franzén, C. J., Taherzadeh, M. J. *J. Biotechnol.* 2011, 156, 22-29.

[5] Kurayama, F., Yoshikawa, T., Yamada, H., Furusawa, T., Sato, M., Suzuki, N. *J. Soc. Powder Technol. Jpn*. 2010, 47, 594-599.

[6] Brady, D., Jordaan, J. *Biotechnol. Lett*. 2009, 31, 1639-1650.

[7] Glinel, K., Sukhorukov, G. B., Möhwald, H., Khrenov, V., Tauer, K. *Macromol. Chem. Phys*. 2003, 204, 1784-1790

[8] Gong, X. C., Luo, G. S., Yang, W. W., Wu, F. Y. *Sep. and Purif. Technol.* 2006, 48, 235-243.

[9] Vergaro, V., Scarlino, F., Bellomo, C., Rinaldi, R., Vergara, D., Maffia, M., Baldassarre, F., Giannelli, G., Zhang, X., Lvov, Y. M., Leporatti, S. *Adv. Drug Delivery Rev*. 2011, 63, 847-864.

[10] Park, J. K., Hang, H. N. *Biotechnol. Adv*. 2000, 303-319

[11] Spiekermann, P.,Vorlop, K.D., Klein, J. *Proceedings of* 4th *European Congress on Biotechnology* 1987, Vol.3, 590-593.

[12] Sasaki, E., Kurayama, F., Ida, J., Matsuyama, T., Yamamoto, H. *J. Electrostat.* 2008, 66, 321-318

[13] Koyama, K., Seki, M. *J. Biosci. Bioeng*. 2004, 98, 114-121.

[14] Tanriseven, A., Dogan, S. *Process Biochem.* 2001, 36, 1081-1083.

[15] Blandino, A., Macías, M., Cantero, D. *Process Biochem*. 2001, 36, 601-606.

[16] Jiang, Z., Zhang, Y., Li, J., Jiang, W., Yang, D., Wu, H. *Ind. Eng. Chem. Res.* 2007, 46, 1883-1890.

[17] Blandino, A.,Macías, M.,Cantero, D. *Appl. Biochem. Biotechnol*. 2003, 110, 53-60.

[18] Catana, R., Ferreira, B. B., Cabral, J. M. S., Fernandes, P. *Food Chem.* 2005, 91, 517-520.

[19] Srivastava, R., Brown, J. Q., Zhu, H., McShane, M. J. *Macromol. Biosci*. 2005, 5, 717-727.

[20] Coradin, T., Nassif, N., Livage, J. *Appl. Microbiol. Biotechnol*. 2003, 61, 429-434.

[21] Zhang, L., Jiang, Y., Jiang, Z., Sun, X., Shi, J., Li, L., Li, J. *J.Biomater. Sci. Polym.* Ed. 2009, 20, 1661-1674.

[22] Jiang, Y., Zhang, L., Yang, D., Li, L., Zhang, Y., Li, J., Jiang, Z. *Ind. Eng.Chem. Res.* 2008, 47, 2495-2501.

[23] Zhang, L., Jiang, Y., Shi, J., Sun, X., Li, J., Jiang, Z. *React. Funct. Polym.* 2008, 68, 1507-1515.

[24] Sakai, S., Ono, T., Ijima, H., Kawakami, K. *Biomaterials* 2002, 23, 4177-4183.

[25] Lim, S. Y., Kim, K. O., Kim, D. M., Park, C. B. *J. Biotechnol*. 2009, 143, 183-189.

[26] Hwang, E. T., Lee, H., Kim, J. H., Tatavarty, R., Gu, M. B. *J. Mater. Chem.* 2011, 21, 6491-6493.

[27] Boninsegna, S., Toso, R. D., Monte, R. D. *J. Sol-Gel Sci. Technol.* 2003, 26, 1151-1157.

[28] Coradin T., Mercey, E., Lisnard, L., Livage, J. *Chem. Commun.* 2001, 2496-2497.

[29] Zhang, Y., Wu, H., Li, J., Li, L., Jiang, Y., Jiang, Y., Jiang, Z. *Chem. Mater*. 2008, 20, 1041-1048.

[30] Zhang, Y., Wu, H., Li, L., Li, J., Jiang, Z., Jiang, Y., Chen, Y. *J. Mol. Catal. B: Enzym*. 2009, 57, 130-135.

[31] Boissiére, M., Meadows, P. J., Brayner, R., Hélary, C., Livage, J., Coradin, T. *J. Mater. Chem*. 2006, 16, 1178-1182.

[32] Golub, A. A., Zubenko, A. I.,Zhmud, B. V. J. *Colloid Interface Sci.* 1996, 179, 482-487.

[33] Zou, H., Wu, S., Shen, J. *Chem. Rev.* 2008, 108, 3893-3957.

[34] Caruso, F. *Adv. Mater.* 2001, 13, 11-22.

[35] Beari, F., Brand, M., Jenkner, P., Lehnert, R., Metternich, J. H., Monkiewicz, J., Siesler, H. W. *J. Organomet. Chem.* 2001, 625, 208-216.

[36] Trens, P., Denoyel, R., Rouquerol, J. *Langmuir* 1995, 11, 551-554.

[37] Plueddemann, E. P. *Silane Coupling Agents,* second ed., NewYork and London, Plenum Press, 1991.

[38] Kurayama, F., Suzuki, S., Oyamada, T., Furusawa, T., Sato, M., Suzuki, N. *J. Colloid Interface Sci.* 2010, 349, 70-76.

[39] Tishkov, V. I., Popov, V. O. *Biomol. Eng.* 2006, 23, 89-110.

[40] Bradford, M. *Anal. Biochem.* 1976, 72, 248-254.

[41] Ma, L., Lu, W., Wen, J. *J. Mol. Catal. B: Enzym.* 2009, 56, 102-107.

[42] Thu, B., Bruheim, P., Espevik, T., Smidsrod, O., Soon-Shiong, P., Skjåk-Bræk, G. *Biomaterials* 1996, 17, 1069-1079.

[43] Zhu, H., Srivastava, R., McShane, M. J. *Biomacromolecules* 2005, 6, 2221-2228.

[44] Xu, S. W., Jiang, Z. Y., Lu, Y., Wu, H., Yuan, W. K. *Ind. Eng. Chem. Res.* 2006, 45, 511-517.

[45] Sakai, S., Ono, T., Ijima, H., Kawakami, K. *J. Sol-Gel Sci. Technol.* 2003, 28, 267-272.

[46] Naviroj, S., Culler, S. R., Koenig, J. L., Ishida, H. *J. Colloid Interface* Sci. 1984, 97, 308-317.

[47] Zhang, L., Liu, J., Yang, J., Yang, Q., Li, C. *Microporous Mesoporous Mater.* 2008, 109, 172-183.

[48] Chong, A. S. M., Zhao, X. S. *J. Phys. Chem.* 2003, 107, 12650-12657.

[49] Valentin, R., Horga, R., Bonelli, B., Garrone, E., Renzo, F. D., Quignard, F. *Biomacromolecules* 2006, 7, 877-882.

[50] Wu, H., Zhao, Y., Nie, M., Jiang, Z. *Sep. Purif. Technol.* 2009, 68, 97-104.

[51] Tanaka, T., Fillmore, D. J. *J. Chem. Phys.* 1979, 70, 1214-1218.

[52] Torres, E., Nata, Y. N., Blázquez, M. L., Muñoz, J. A., González, F., Ballester, A. *Langmuir* 2005, 21, 7951-7958.

[53] Etienne, M., Walcarius, A. *Talanta* 2003, 59, 1173-1188.

[54] Lewińska, D., Rosiński, S., Hunkeler, D., Poncelet, D., Weryński, A. *J. Membr. Sci.* 2002, 209, 533-540.

[55] Chai, Y., Mei, L. H., Lin, D. Q., Yao, S. J. *J. Chem. Eng. Data* 2004, 49, 475-478.

[56] Chai, Y., Mei, L. H., Wu, G. L., Lin, D. Q., Yao, S. *J. Biotechnol. Bioeng.* 2004, 87, 228-233.

[57] Tomida, H., Nakamura, C., Yoshitomi, H., Kiryu, S. *Chem. Pharm. Bull.* 1993, 41, 2161-2165.

[58] Kaplan F., Setlow P., Kaplan N.O. *Arch. Biochem. Biophys.* 1969, 132, 91-98.

[59] Vincent, K. A., Parkin, A., Armstrong, F. A. *Chem. Rev.* 2007, 107, 4366-4413.

[60] Liu, W., Wang, P. *Biotechnol. Adv.* 2007, 25, 369-384.

[61] Migneault, I., Dartiguenave, C., Bertrand, M. J., Waldron, K. C. *BioTechniques* 2004, 37, 790-802.

[62] Jayakrishnan, A., Jameela, S. R. *Biomaterials* 1996, 17, 471-484.

[63] Antipov, A. A., Sukhorukov, G. B., Leporatti, S., Radtchenko, I. L., Donath, E., Möhwald, H. *Colloids Surf.*, A 2002, 198-200, 535-541.

[64] Kawaguti, H. Y., Buzzato, M. F., Orsi, D. C., Suzuki, G. T., Sato, H. H. *Process Biochem.* 2006, 41, 2035-2040.

[65] Wang, Z., Qian, L., Wang, X., Zhu, H., Yang, F., Yang, X. *Colloids Surf.,* A 2009, 232, 164-171.

[66] Schütte, H., Flossdorf, J., Sahm, H., Kula, M. R. *Eur. J. Biochem.* 1976, 62, 151-160.

In: Silane: Chemistry, Applications and Performance
Editors: Katsuo Moriguchi and Susumu Utagawa
ISBN: 978-1-62257-432-2

Chapter 7

SYNTHESIS OF A MESOPOROUS SILICA HOLLOW MICROSPHERE USING METHYLTRIMETHOXY SILANE

N. Venkatathri[*]
Department of Chemistry, Anna University, Chennai, Tamil Nadu, India

ABSTRACT

A mesoporous silica hollow micro sphere was synthesized in presence of methyltrimethoxy silane for the first time. The synthesized material was characterized by various instrumental techniques such as small-angle X-ray scattering, scanning electron microscopy, transmission electron microscopy.

Thermogravimetry/Differential thermal analysis, Fourier transform Infrared spectroscopy and Nitrogen adsorption measurement. The analysis reveals the material is mesoporous and having hollow micosphere morphology.

Keywords*:* Silica; hollow micro sphere; methyltrimethoxy silane

1. INTRODUCTION

There is a great interest concerning the synthesis of hollow materials with specific structure and different compositions [1-4]. It is due to their applications in catalysis, controlled release for drugs, dyes and perfumes, development of artificial cells and protection of biologically active agents such as proteins and enzymes [5, 6]. These uses are due to its properties of lower densities, high surface area and more unique optical and chemical properties. Several techniques reported for the synthesize of this materials such as nozzle

[*] E-mail address: venkatathrin@yahoo.com. Tel.: +91-9486565702.

reactor approaches, emulsion/phase extraction techniques and self assembly processes including layer by layer (LBL) method [7,8].

In this paper, we present a simple pathway to fabricate silica hollow micro sphere. The synthesized material was characterized by different physicochemical techniques.

2. Experimental

6 ml of tetraethyl orthosilicate (98 %, Aldrich, USA) was mixed with 3.48 ml of methyl trimethoxy silane (98 %, Aldrich, USA). 1.50 ml of distilled water was mixed with the above solution. The mixture was stirred for 5 min. 1.98 ml of aqueous ammonia (18-30 %, Aldrich, USA) was mixed with 70 ml of distilled water and emulsified for another 10 min. The above silica solution (4.98 ml) was mixed with ammonia solution (71.98 ml) and homogenized at 1000 rpm for another 10 min. 0.836 g of polyvinyl alcohol (Mn 31,000, Aldrich, USA) was mixed with the above mixture and stirred for another 2 h. The final residue was centrifuged to separate and dried at 60°C.

All the samples were analyzed by using the following instruments. Small-angle X-ray scattering (SAXS) were recorded on Rigaku Multiplex diffractometer using Cu Kα radiation and a proportional counter as detector. A divergence slit of 1/328 on the primary optics and an anti-scatter slit of 1/168 on the secondary optics were employed to measure data in the low angle region.

The scanning electron micrographs were taken using a Topcon, SM-300. The samples were deposited on a sample holder with an adhesive carbon foil and sputtered with gold. For transmission electron microscopy, the samples were ground and deposited on a circular disc of fine copper mesh covered with collodion. Images were taken using a JEOL JSM-2000 EX electron microscope operated at an acceleration voltage of 200 kV. Simultaneous TG/DTA analyses of the crystalline phase were performed on an automatic derivatograph (Setaram TG-DTA 92).

The specific surface area (BET) of the samples was determined using a Micromeritics ASAP 2010 volumetric adsorption analyzer. Before N_2 adsorption samples was evacuated in vacuum at 300°C. The data points of p/p_0 in the range of about 0.05–0.3 were used in the calculations. The FT-IR spectra in the framework region were recorded in the diffuse reflectance mode (Nicolet 60SXB) using 1:300 ratio of sample with KBr pellet.

3. Results and Discussion

Figure 1 illustrates the small-angle x-ray scattering (SAXS) patterns of the as-synthesized sample. The single and broad peak around 1.5° in 2 theta shows a mesoporous structure. The peak also did not change on calcinations at 550°C for removal of the entrapped organics inside the pores. However there is slight reduction in intensity is observed. Scanning electron microscopic images of the samples synthesized from Methyltrimethoxy silane shows spherical shape (Figure 2, 500 nm-3μm).

The smooth internal structure of the sample can be seen from the picture. The white outer layer with dark inner layer reveals its hollow nature.

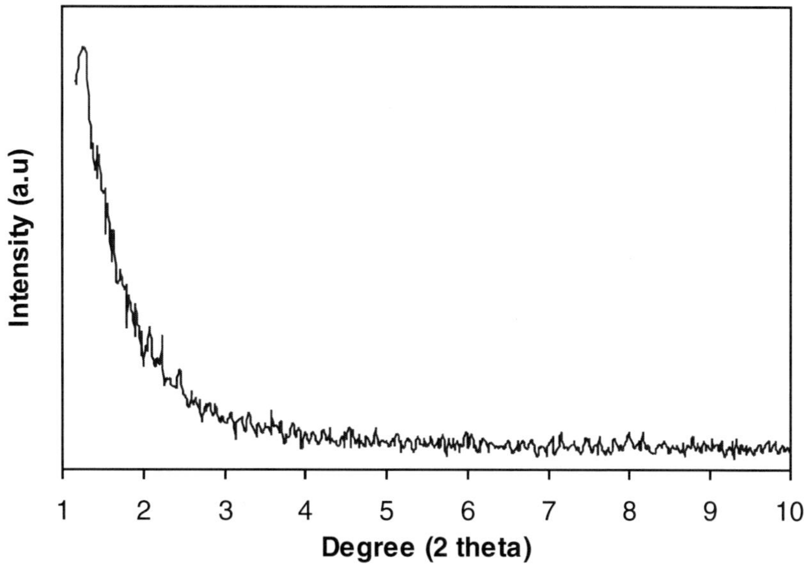

Figure 1. Small angle X-ray diffraction pattern of the mesoporous silica hollow micro sphere.

Figure 2. Scanning electron micrograph of the mesoporous silica hollow micro sphere.

Further the sample's Transmission electron microscope picture (Figure 3) shows (one of the hollow sphere is arrow marked) the hollow sphere property, with 500 nm diameter and 50 nm shell thickness. The Thermogravimetry/Differential thermal analysis curve of synthesized sample is given in Figure 4. The sample losses the template in three stages (7.2 % (25-418 °C), 2.3 % (418-444 °C) and 8.3 % (444-650 °C). The initial endothermic loss is due to the loss of physisorbed material. The latter two exothermic losses are due to oxidative decomposition of the occluded organic material. Totally it losses 21.23 % (25 – 900 °C). Here polyvinyl alcohol functions as template, so the loss due to it is less compared to surfactants like cetyltrimethylammonium bromide (40 %).

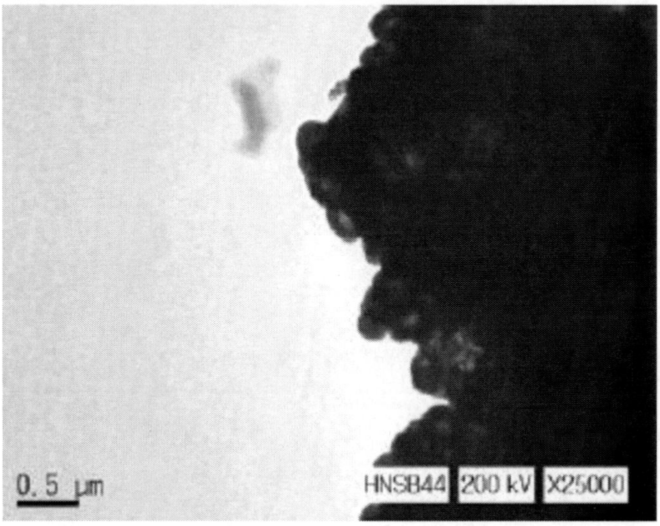

Figure 3. Transmission electron micrograph of the mesoporous silica hollow micro sphere.

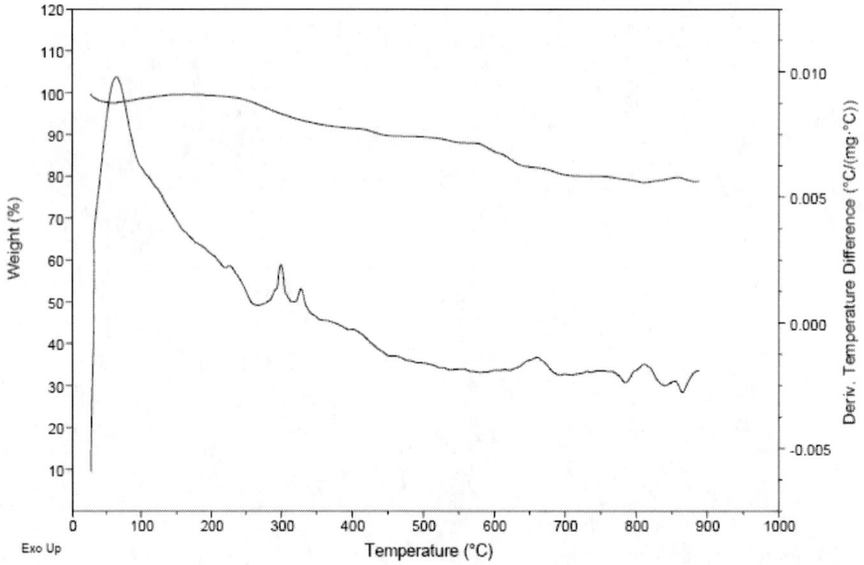

Figure 4. Thermogravimetry/Differential thermal analysis of the mesoporous silica hollow micro sphere.

The Nitrogen adsorption isotherms of silica hollow micro sphere (Figure 5) are characteristics of mesoporous materials with uniform pore size [9]. These isotherms show an inflection near P/Po 0.72 - 0.96, indicting capillary condensation within the pores.

The presence of hysteresis, characteristic of multilayer adsorption dominating the process of filling and emptying the voids, indicates the pore size [10].

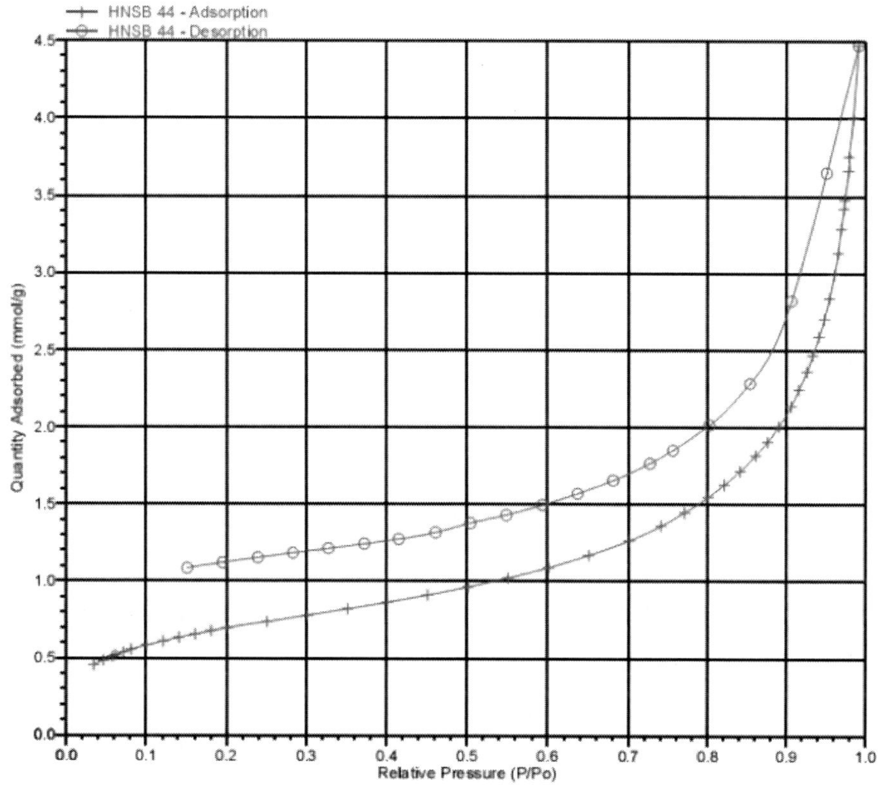

Figure 5. Nitrogen adsorption analysis of the mesoporous silica hollow micro sphere.

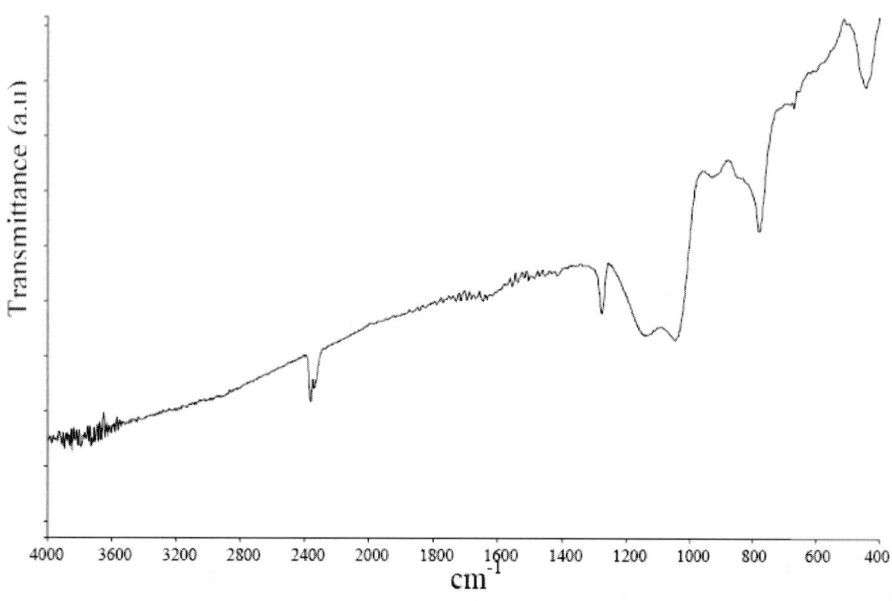

Figure 6. Fourier transform Infrared spectroscopic analysis of mesoporous silica hollow micro sphere.

The BET surface area, is 300 m^2/g, the average pore diameters is 119 Å and cumulative adsorption pore volume (155 cc/g) as calculated using the BJH [11] model, Horvath-Kawazoe medium pore width 196 A^o. The amount of mesoporous area is lesser compared to other reported mesoporous molecular sieves, however comparable to other hollow silica spheres [12]. The Fourier transform Infrared spectroscopy of as-synthesized sample shows (Figure 6) peaks around 1700 and 3430 cm^{-1} corresponding to the carboxyl and hydroxyl groups [13] respectively. Carboxyl peak appeared from the atmospheric carbon dioxide.

The adsorption peak belonging to the Si-O stretching vibration of Si-OH bond appears at 960 cm^{-1}[14]. The weak peaks at 2855 and 2920 cm^{-1} belong to the stretching vibrations of C-H bonds, which show a few organic groups are adsorbed on the spheres. The strong peaks near 1100, 802 and 467 cm^{-1} agree to the Si-O-Si bond which implies the condensation of silicon source [15]. A plausible mechanism of silica hollow microsphere formation is as follows. Initially polyvinyl alcohol spheres were templated to form silica hollow micro sphere. Tetraethylorthosilicate hydrolyzed to hydroxides and precipitated around the polyvinyl alcohol. Ammonia and Methyl trimethoxy silane were acted as catalyst and initiator respectively. Polyvinyl alcohol also functions as dispersant.

Conclusion

Silica hollow microsphere was synthesized from Methyltrimethoxy silane. The synthesized material was characterized by various physicochemical techniques such as small-angle X-ray scattering, Scanning electron microscopy, Transmission electron microscopy, Thermogravimetry/differential thermal analysis, Fourier transform Infrared spectroscopy and Nitrogen adsorption. The results show the material is mesoporous and hollow micro sphere property.

Acknowledgment

The author N.V thanks Council of Scientific and Industrial Research, New Delhi, India, for a Senior Research Associateship.

References

[1] Y. Xiong, Y. Xie, Z. Li, C. Wo, R. Zhang, *Chem. Commun.* 7 (2003) 904.

[2] K.J.C. van Bommel, J.H. Jung, S. Shinkai, *Adv. Mater.* 13 (2001) 1472.

[3] T. Kato, H. Ushijima, M. Katsumata, T. Hyodo, Y. Shimizu, M. Egashira, *J. Mater. Sci.* 37 (2002) 2317.

[4] J.J. Zhu, S. Xu, H. Wang, J.M. Zhu, H.Y. *Chen, Adv. Mater.* 15 (2003) 156.

[5] E. Mathlowitz, J.S. Jacob, Y.S. Jong, G.P. Carino, D.E. Chickering, P. Chaturvedi, C.A. Santos, K. Vijayaraghavan, S. Montgomery, M. Bassett, C. Morrell, *Nature* 386 (1997) 410.

[6] H. Huang, E.E. Remsen, T. Kowalewski, K.L. Wooley, *J. Am. Chem. Soc.* 121 (1999) 3805.

[7] P.D. Yang, D.Y. Zhao, B.F. Chmelka, G.D. Stucky, *Chem. Mater.* 10 (1998) 1033.

[8] F. Caruso, R.A. Caruso, H. Mohwald, *Science* 282 (1998) 1111.

[9] S. Brunauer, L.S. Deming, W.S. Deming, E. Teller, *J. Am. Chem. Soc.* 62 (1940) 1723.

[10] J.H. de Boer, *The Structure and Properties of Porous Materials*, Butterworths, London, (1958).

[11] IUPAC, Reporting Physisorption Data for Gas/Solid Systems, *Pure Appl. Chem.* 87 (1957) 603.

[12] W. Fan, L. Gao, J. *Colloid Interface Science* 297 (2006) 157.

[13] Y. H. Li, C. Xu, B. Wei, X. Zhang, M. Zheng, D. Wu, P.M. Ajayan, *Chem. Mater.* 14 (2002) 483.

[14] Y. Shan, L. Gao, S. Zheng, *Mater. Chem. Phys.* 88 (2004) 192.

[15] J.R. Agger, M.W. Anderson, M.E. Pemble, O. Terasaki, Y. Nozue, *J. Phys. Chem. B,* 102 (1998) 3345.

INDEX

A

B

C

D

E

F

G

H

I

J

K

L

M

N

O

P

T